AN INTRODUCTION TO SPECIALITY POLYMERS

AN INTRODUCTION TO

Speciality Polymers

Edited by

NORIO ISE & IWAO TABUSHI

Kyoto University, Japan

CAMBRIDGE UNIVERSITY PRESS

Cambridge

London New York New Rochelle

Melbourne Sydney

Published by the Press Syndicate of the University of Cambridge
The Pitt Building, Trumpington Street, Cambridge CB2 1RP
32 East 57th Street, New York, NY 10022, USA
296 Beaconsfield Parade, Middle Park, Melbourne 3206, Australia

Originally published in Japanese by Iwanami Shoten, Tokyo, 1980 and
© Norio Ise and Iwao Tabushi, 1980

First published in English by Cambridge University Press 1983
as *An Introduction to Speciality Polymers*
English edition © Cambridge University Press 1983

Printed in Great Britain at the University Press, Cambridge

Library of Congress catalogue card number: 82-22012

British Library cataloguing in publication data
Ise, N.
An introduction to speciality polymers.
1. Polymers and polymerization
I. Title   II. Tabushi, I.
547.7   QD381

ISBN 0 521 24536 2

DJ

CONTENTS

CONTRIBUTORS

Morio Ikehara
Department of Pharmaceutical Sciences, Osaka University, Osaka, Japan

Norio Ise
Department of Polymer Chemistry, Kyoto University, Kyoto, Japan

Toyoki Kunitake
Department of Organic Synthesis, Kyushu University, Fukuoka, Japan

Makoto Okawara
Research Laboratory of Resources Utilization, Tokyo Institute of Technology, Yokohama, Japan

Manabu Senō
Institute of Industrial Sciences, University of Tokyo, Tokyo, Japan

Iwao Tabushi
Department of Synthetic Chemistry, Kyoto University, Kyoto, Japan

Shigeo Tazuke
Research Laboratory of Resources Utilization, Tokyo Institute of Technology, Yokohama, Japan

PREFACE

Keen attention is currently being paid to the field of speciality polymers. The motivation is our naïve admiration of naturally occurring biopolymers, which have highly sophisticated functions, and also our dissatisfaction with the roles which we have been giving to synthetic polymers. To what extent can we mimic natural compounds by using synthetic ones? This question is not at all novel: scientists, in particular synthetic organic chemists, have continually addressed it. Thus interest in speciality polymers existed already and did not start abruptly. The answer to the question is one of the goals that macromolecular chemistry should ultimately aim to reach.

In this book, some selected, recent topics on speciality polymers will be discussed. It must be mentioned that we did not intend to cover all topics thoroughly in this area, mainly because of space limitation.

Following some introductory remarks in Chapter 1, Chapter 2 describes the synthetic chemistry of speciality polymers. Though research activity on speciality polymers is rather new, the desired polymers cannot be prepared without classical organic chemistry. It is obviously advantageous to employ widely used, familiar macromolecules as starting materials. Thus the molecular structures of the product polymers cannot be so very different from those familiar to synthetic polymer chemists; consequently the functions of the product polymers are in many cases far removed from those exhibited by naturally occurring biopolymers. However, what has been achieved so far marks a step taken forward from the traditional way of thinking which pointed solely to the mechanical properties of macromolecules.

The highly coordinated and efficient functions of living organisms are supported by the biocatalysts, namely enzymes. Their remarkable functions attract not only biochemists but also polymer chemists who have recently started preparing synthetic macromolecules having catalytic activity. Some of them are studying intentionally very simple reactions, and likewise simple synthetic macromolecules, in order to obtain the most basic information about enzymic catalysis (Chapter 3, part A). Simultaneously, others are preparing new macromolecular catalysts by introducing catalytically active groups and studying their functions (Chapter 3, part B). It is true that synthetic compounds may be superior to enzymes in certain of their properties. The mechanism by which living systems carry out photosynthesis, harnessing light energy in a most efficient manner, teaches us an important lesson. The energy–matter interconversion process must be investigated urgently from both basic and practical standpoints. The role which natural macromolecules are playing in the process of energy conversion is also great so that intensive effort is being made to study whether it is possible to replace them by synthetic macromolecules, as is discussed in Chapter 4. It is inferred that the substrate specificity and stereospecificity often encountered in biological systems has some bearing with transport, membrane and interfacial phenomena in which macromolecules are involved. The quantitative study of these phenomena is definitely desirable; the current status of our knowledge on this subject is discussed in Chapter 5. Chapter 6 is devoted to a discussion of recent advancement in the study of synthetic membranes. This new-born field should contribute greatly to our understanding of the properties and functions of biological membranes. In Chapter 7, macromolecules that convey information, namely nucleic acids, and their model compounds are treated. The diversity of the information and its exact transfer are absolutely necessary for living organisms. As is well-known, the polynucleotides are copolymers of four different kinds of monomers with well-defined orders. On the other hand, we polymer chemists can barely prepare alternate copolymers from two monomers. Obviously the gulf between polynucleotides and synthetic macromolecules appears to be hopelessly large. For this reason, we may say that many lessons are yet to be learned in this complicated field.

We are eye-witnesses of the remarkable advances in other fields such as polymeric drugs which are not discussed in this book. But even including these topics, it is still premature to claim that significant achievements have been made in the study of speciality polymers. Probably, functions expected now for speciality polymers are much more diversified and intricate, requiring delicate balances of many factors, than those polymer chemists have successfully provided with the conventional synthetic macromolecules. It is evident that intensive interdisciplinary study must be carried out to achieve this goal.

Finally a few words appear to be necessary for non-Japanese readers. This book is an English version of the original Japanese one, which was published in February, 1980 by Iwanami-Shoten, Tokyo, as one of a series entitled *Modern Chemistry*. In order to avoid repetition, many subjects, even though pertinent to speciality polymers, were omitted in this book, but are discussed in others in the series. For example, a detailed description of enzyme mechanism is totally ignored, although it is a topic closely related to macromolecular catalysis, which is discussed in depth in this book. Second, the original version was written for Japanese undergraduate and graduate students. Partly for this reason, emphasis was put sometimes – though not always, – on publications that are easily accessible to Japanese readers. Third, in light of the rapid advancement in the field concerned, new results were incorporated in the English version when it appeared necessary. Furthermore, the editors invited Professor Toyoki Kunitake to contri-

bute a new chapter on the synthetic membrane system, a field which became
especially interesting after the original manuscript for the Japanese version had
been prepared. The editors are grateful to all the contributors for their time
spent in preparing the manuscripts and translating them into English. Last, but
not least, thanks are due to Drs E. Kirkwood and A. J. Colborne at the Cambridge
University Press, for the kind editorial assistance in producing the English version
of this book.

1

INTRODUCTION

NORIO ISE

The term 'speciality polymers' is not so well-defined as most other terms in
chemistry. If we understand it in a broad sense, all naturally-occurring and
synthetic macromolecules fall into this category. Some macromolecules have
been necessary for the regulation of body temperature and protection of the
body for a very long time. The basic properties required for these functions
are, above all, relatively low thermal conductance and high mechanical strength.
Without precise tools for measuring these properties, their importance has
been recognized simply from experience. These days, however, synthetic macromolecules have appeared on the scene in addition to the natural ones, and they
have become more and more frequently used for this purpose. In some cases,
synthetic macromolecules performed much better than natural ones, so that
the latter have lost their exalted position, at least partly. Nowadays, we find
numerous plastics and fibres around us, which were produced from synthetic
macromolecules: containers from polypropylene, coating materials from
poly(vinyl chloride), experimental apparatus from Teflon, organic glasses from
poly(methyl methacrylate), etc. – there are simply too many to mention them
all. The reason why these materials are popular is that they have satisfactory
stability, electrical insulation, mechanical strength, and so on.

Thus, our daily life cannot be without these macromolecular substances,
whether we like it or not. However, if we look at the situation more closely, we
immediately notice that our way of utilization of the macromolecules is
unbalanced. In other words, the excellent mechanical strength, stability and
so on are only parts of the whole properties which the macromolecules may
be expected to have. It is true that they have been used in a proper manner
according to our necessities, but other ways of utilizing both natural and
synthetic macromolecules may remain. In order to investigate these further
possibilities, active research has been initiated in a field called 'speciality
polymers'.

In our efforts to search for this hidden potential, our compass is the various
functions of living organisms; it is obvious that macromolecules play important
roles in biological functions. Also, we know that the macromolecules that are
synthesized in living organisms – or biopolymers – have distinctive structures.
We can appreciate readily that these functions and structures of the biopolymers

are quite different from what we polymer scientists have experienced with synthetic macromolecules.

Let us think about two examples. Firstly, we consider enzyme molecules, which are well known to demonstrate most admirable catalytic functions in living organisms. The enzymes are copolymers of amino acids. Chymotrypsin is composed of about 250 amino acid residues. However, the important point to note is that a well-defined number of different amino acids are bound to each other in a definite order. On the other hand, synthetic macromolecules have a broad distribution in the degree of polymerization (and hence of the molecular weight), and furthermore most of them are composed of a single kind of monomer; in other words, they are homopolymers. Furthermore, the monomer sequence of synthetic copolymers is usually at random, with exceptional cases such as alternating copolymers of two monomers. Returning to chymotrypsin, it has to be pointed out that its extraordinary catalysis is based ultimately on the well-defined amino acid sequence. Simply for this reason, the molecule assumes a definite spatial shape and an extension, which make possible the binding with substrates of a specified orientation. Furthermore, the ordered sequence makes concerted action among functional groups of an enzyme molecule possible. The following scheme (after M. Bender) shows the deacylation process of chymotrypsin, in which the enzyme molecule is depicted as a line

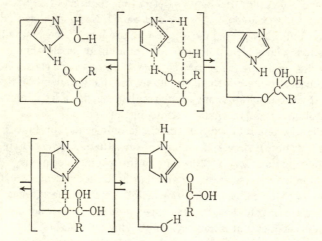

connecting imidazolyl and serine groups; the imidazolyl group acts as a general base to cut off an acyl group bound on the hydroxyl group of serine. Obviously, this concerted mechanism is possible only when the two groups are close to each other. This condition is fulfilled as a consequence of the characteristic molecular geometry, which in turn originates from the ordered sequence of the amino acids.

Secondly, we need to mention the stereospecific structure of biopolymers. Cellulose and starch, having the same chemical composition, display completely different roles. Most surprising is the fact that the stereospecific polymerization for each of them is carried out without error in living systems. In the case of vinyl polymerization, on the other hand, the stereospecific arrangement in the neighbourhood of α-carbons being produced along the polymer chain can be controlled to some extent by using, for example, a $TiCl_3$–$Al(C_2H_5)_3$ catalyst system. It has been reported that polypropylene can be prepared having a 85% isotacticity. However, the 100% stereospecificity, which is usual for the biopolymers, has not been generally realized for synthetic macromolecules.

It is not clear how these strict regulations are made possible in living systems; thus it is no wonder if a strong desire emerges to solve the mystery. Furthermore, various functions in living systems are highly attractive to polymer scientists, who may make efforts to mimic naturally occurring macromolecules by synthetic ones, namely speciality polymers.

If we regard synthetic macromolecules used for purposes other than fibres and plastic products as forerunners of speciality polymers, the study of speciality polymers may be claimed to have started a fairly long time ago. One example is ion-exchange resins. They absorb selectively cations or anions. Thanks to these polymeric products, the purification of water was greatly facilitated. Shortly thereafter, we realized that ion-exchanged water was pure ionically but non-ionic impurities could not be removed. This fact appears to demonstrate the limitations of speciality polymers. It can be easily understood from this book that synthetic macromolecules can demonstrate fairly complicated (in many cases, a single) functions, but at the same time it must be admitted that this is only a tiny part of a number of functions of the biopolymers. For example. enzymes have (i) an excellent catalytic activity, (ii) a rather strict substrate specificity and (iii) a reaction-control function. It is not an exaggeration to state that synthetic macromolecules can imitate fairly well only the catalytic activity of enzymes. They barely show specificity and are completely powerless as far as the reaction-control function is concerned. This difference is due to the fact that enzymes are the product of evolution in a time span of over a thousand million years, whereas only fifty years have elapsed since the birth of synthetic macromolecules. To what extent the gap between the natural biopolymers and speciality polymers can be made narrower by future research is most interesting to contemplate.

2

REACTIVE POLYMERS

MAKOTO OKAWARA

2.1   **What are reactive polymers?**

As is well known, many high polymers have sufficient strength and stability for practical use as materials. These desirable properties permit us to use them for long periods. This is a reason why synthetic polymers have been widely used in our daily life ever since they were first synthesized. For example, heat-resistant polymers have been developed for the purpose of acquiring heat resistance without deformation at temperatures as high as several hundreds of degrees. Plastics have been widely applied as construction materials or automobile parts. Nylon fibre has been used as a material for stockings. The widely produced polymers, so-called industrial plastics, fibres and rubbers have reached the stage that they cannot be improved further in their intrinsic structure. It is also difficult to design any organic polymers that are simpler in structure and can be manufactured at a lower price than polyethylene $-(CH_2-CH_2)_n$. The annual production of plastics in Japan has reached several million tons in the last few years. Many problems concerning consumption of resources and environmental pollution (besides treatment of waste materials) have arisen together with increasing production of polymers. Thus, further development of polymers is now at a turning point.

Previous investigations aimed at developing inexpensive and strong polymers; in contrast, the synthesis of new polymers with high additional value and specific functions is attracting keen attention. Chemistry relating to speciality polymers, the main subject of this book, is thought to be a new field in polymer science; this field is now starting to develop. In order to provide an additional function to polymers, e.g. electrical conductivity or photosensitivity, it is necessary that the corresponding functional group (F) be introduced into the polymers. It is not easy, however, to introduce directly such specific F into stable polymeric materials. By analogy with the concept of reactive intermediates in organic synthesis, reactive polymers are considered to be intermediates (P—R) through which final polymers with the desired function (P—F) are obtained from the stable starting polymers (P).

Reactive polymers are also sought for their own sake in many cases. For example, ion-exchange resins are classified as one speciality polymer possessing a new function (ion-exchange function) which was not found in the original polymers. The ion-exchange resins can be regarded not only as end materials, but they can also be used as reactive polymers. Reactive polymers that form graft polymers are thought to be end products, but they are also regarded as reactive intermediates for graft polymers. Thus, classification of reactive polymers has not been well-defined. For convenience, reactive polymers are classified in broad terms into two types; typical examples are shown in Table 2.1.

Table 2.1. *Reactive polymers*

1. As reactive intermediates:
    main chain polymers for graft- and block-polymers, prepolymers for cross-linkage, reactive polymers for introducing functional groups into polymers, etc.
2. As end polymers:
    ion-exchange resins, chelating polymers, photosensitive polymers, polymer reagents, polymer catalysts, reactive fibres, degradable polymers, etc.

Examples of these are individually discussed in other chapters of this book. In this chapter, some speciality polymers are described from the viewpoint of general applications of reactive polymers. We discuss here the problems concerned with reactive polymers that are regarded as intermediates.

Let us consider the synthesis of reactive polymers. There are two routes to synthesize reactive polymers (P—R) which contain corresponding reactive groups (R). One is polymer reactions; in the other route, P—R is obtained by polymerization or condensation of corresponding monomers which originally include a reactive group (M—R). As explained above, polymers with intended functional structure (P—F) can be conveniently derived from highly reactive P—R. It is also possible to derive P—F by polymerization or condensation of monomers possessing the intended functional structure (M—F) which have been derived from M—R. These relations are illustrated below.

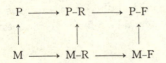

Reactive groups (R) can be combined with the chain of the molecule in the following ways. In (A), a large number of R are regularly bonded in the molecule. In (B) and (C), a small number of R are bonded at random in and at the terminal of the molecule, respectively. Case (A) is generally found in the polymerization of M—R, while case (B) occurs mainly in the product of polymer reactions. Those characteristics and problems will now be explained in detail.

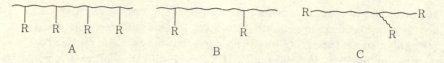

The reactivity of R cannot be generalized simply. Some groups which usually stable polymers contain, such as $-CH_2CH_2-$ in polyethylene and $-O-$ in polyether, exhibit extremely low reactivity. On the other hand, groups like $-OH$ contained in poly(vinyl alcohol) and cellulose, $-CONH-$ in nylon, and $-COOH$ in poly(acrylic acid), are quite reactive in some reactions. However, they are not reactive enough to be used as so-called reactive intermediates. How can their activity be further enhanced? Suppose that reactive functional groups such as epoxide, vinyl acid anhydride, aldehyde, *N*-methylol, chloromethyl, amino or other functional groups are introduced into the original polymer. The obtained polymer will then become one of the typical reactive polymers used in industry. The polymers are also stable. Although polymers which contain such functional groups as isocyanate, acid chloride, thiol and aziridine ring, or which have the structure of a diazonium salt, are certainly reactive, they are too unstable to be preserved. They must therefore be prepared just prior to the reaction, or stored in the form of proper precursors (protected form) which can be reactivated when the reaction starts.

The reactivity of functional groups is also dependent on how they are combined. For example, hydroxyl groups (OH) in cellulose are not reactive enough because of hydrogen bonds; this is why cellulose is insoluble in water. In cellulose modified by partial hydroxyethylation with ethylene oxide, $Cell-O(CH_2CH_2-O)_{1\sim3}-OH$, OH groups are distant from the main chain and become more reactive with enhanced degrees of freedom. Thus, a full understanding of the polymer effect in polymer reaction is required when reactive polymers are designed or are applied in practice. In addition, the choice of polymer chain (polymers as supports for reactive groups) is a key point in the construction of the intended reactive polymers. We shall discuss this in §2.2.2. of this chapter.

Several reviews have been published which deal with reactive polymers (apart from speciality polymers); some of them published in recent years are cited at the end of this book. These reviews should be referred to for details on the reactive polymers.

2.2 Syntheses of reactive polymers
2.2.1 *Classification by reactive group*

As discussed above, there are various reactive groups, which show low, medium or high reactivities. Functional groups of medium and high reactivities are important when high polymers are to be used as intermediates for reactions. Our discussion is here restricted to syntheses and reactivities of polymers with reactive groups in side chains.

(a) Hydroxyl (OH) group. Hydroxyl groups are contained in cellulose, poly(vinyl alcohol) (PVA) and phenol resin. In the first two, the activity of hydroxyl groups is repressed because of strong hydrogen bonds between the hydroxyl groups, as mentioned above. When the crystalline structure is perturbed by modification of some of the hydroxyl groups, the other hydroxyl groups become more reactive. We should not ignore the vinyl acetate monomer. It is not only used as a raw material for PVA, but is also applied to make vinyl alcohol–ethylene copolymers (already commercialized), which possess properties of polyethylene (PE) as well as reactivity of hydroxyl groups to give a practical hot-melt type

$$CH_2=CH \atop CH_2OH \qquad \longrightarrow \quad -CH_2-CH-- \atop CH_2OH \qquad \longleftarrow \quad -CH_2-CH- \atop COOCH_3$$

$$[1] \qquad\qquad\qquad [2]$$

adhesive. Allyl alcohol [1], which is not polymerized to a compound with high molecular weight, is most frequently used in copolymerizations for the purpose of introducing the unit [2], which is available also by reactions of poly(methyl acrylate). More practical uses of the reactivity of OH groups are found in polymers and copolymers of 2-hydroxyethyl methacrylate [3]. Partially

$$-CH_2-C(CH_3)- \atop COOCH_2CH_2OH$$

$$[3]$$

crosslinked polymers and copolymers thereof are hydrophilic as well as insoluble in water; their unique properties are most valuable in applications to contact lenses, etc. Also bromocyanation effectively converts OH groups contained in the polymers of 2-hydroxyethyl methacrylate, cellulose and agarose, to more active groups, as shown in [4]; these groups are suitable for applications in the circumstances analogous to those in living organisms, (as we see from immobilized enzymes). The polymer [5] obtained from vinylene carbonate has

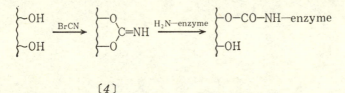

$$[4]$$

a high degree of polymerization, and is hydrolysed to a compound with vicinal hydroxyl groups. It may be expected to have properties different from those of PVA and analogous to those of polysaccharides. It is troublesome, however, that [6] has a limited solubility, owing to the extremely strong hydrogen bonds. OH groups are added to epoxy resins by so-called curing reactions. Copolymers of glycerol (containing three OH groups) with phthalic or maleic acid (contain-

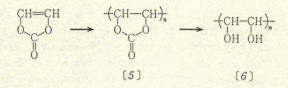

[5] [6]

ing two COOH groups) form a three-dimensional structure with unreacted OH and COOH groups. This expands their applicability, by deformation of OH groups and other treatments, to paints, inks, and other materials (paint bases).

There are various techniques for introduction of OH groups into polymers. For example, diene-type rubbers like polybutadiene contain $-C=C-$ units in the main chain (in the case of 1,4-polymers) or in branches (in the case of 1,2-polymers), where OH groups can be introduced by hydrolysis of epoxide, carbonylation followed by reduction, or hydroboration, and so on. The halogens

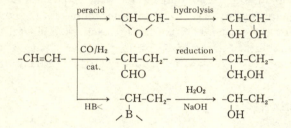

in some polymers can be substituted by OH groups by hydrolysis. However, in order to obtain a reactive polymer with OH groups, hydrolysis of an ester (e.g. acetate) derived as an intermediate is much easier than direct hydrolysis of halogen-containing polymer; this has been proved in the case of chloro-methyl polystyrene [7] and polyepichlorohydrin. The introduction of OH

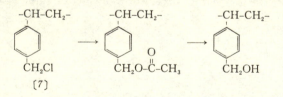

[7]

groups into polymers aims not only at obtaining reactive intermediates but also at improving the properties of the polymers. When a small quantity of OH groups is introduced, hydrophilicity, antistatic properties, and fitness for printing (inks) of PE, PP (polypropylene) and other hydrophobic polymers are improved. Also for these purposes, PE and other polymers are oxidized to contain very small amounts of $-C=O$ or $-OOH$ groups (c. 0.1%), and then reduced to polymers with the desired properties.

The typical polymer containing phenolic hydroxyl groups is undoubtedly phenol–formalin resin. The synthesis of monomers [8], [9] and the homo- and copolymerization thereof have been recently developed in Japan. This acidic OH group, of rather different reactivity from an alcoholic one, is of interest in electrophilic substitution reactions at the *ortho* and *para* positions.

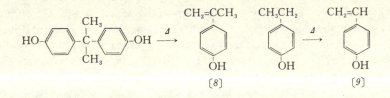

[8]   [9]

Silica and glass fibres, which are representative inorganic polymers, contain acidic OH groups in the form of $-O-Si-OH$ at the surface (in other words, at the terminal), most of the OH groups being too stabilized by hydrogen bonds to employ them in reactions. These materials can be regarded as reactive polymers because research on activation of the OH groups has developed remarkably. These topics will be examined in detail in §2.2.2. of this chapter.

As explained above, OH groups in polymers are too inactive for the polymers to be used as reactive intermediates. Therefore introduction of OH groups aims, in many cases, at enhancement of the hydrophilic properties of polymers. Esterification, etherification, carbamidation, oxidation and other reactions in which active hydrogen participates will be mentioned for PVA and cellulose. We have discussed how OH groups of a low activity are activated to become electrophiles. Then, how are they activated to nucleophiles? An example is given by hydroxyethyl methacrylate ([3], $R = CH_3$). When OH groups in poly-(hydroxymethyl methacrylate) are transformed into tosyl groups [10], the density of electrons becomes remarkably low at the carbon atoms which have combined with strongly electron-withdrawing tosyl groups. Most nucleophiles (e.g. $Nu = RS^{\ominus}$, $CH_3COO^{\ominus}$ and $C_6H_5O^{\ominus}$) can easily attack carbon atoms to produce C-Nu derivatives. Polymers containing hydroquinones, [11], can be classified as OH polymers. Since it is one of the redox-type polymers, it should be regarded as a representative functional polymer.

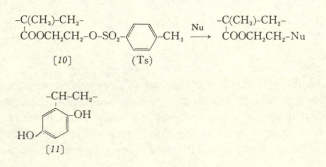

$-C(CH_3)-CH_2-$
$COOCH_2CH_2-O-SO_2-$〈 〉$-CH_3$ $\xrightarrow{Nu}$ $-C(CH_3)-CH_2-$
$COOCH_2CH_2-Nu$

[10]   (Ts)

[11]

(b) Carboxyl (COOH) groups and their derivatives. Homopolymers and copolymers of acrylic and methacrylic acids are representative reactive polymers with COOH groups. Being acidic and hydrophilic (soluble in water), the homopolymers are not fully used as polymeric materials. The copolymers of the acids with divinyl benzene (DVB) are exceptionally insoluble in water, and form metal salts; these copolymers are known as weakly acidic ion-exchange resins. In many cases, these monomers are used as components for copolymerization, and for the purpose of endowing hydrophobic polymers with hydrophilic properties, and they display some ability to crosslink with metal cations (e.g. preparation of ionomers by introducing acrylic or methacrylic acid to PE). Besides direct polymerization of the corresponding monomers, polystyrene with carboxylic groups [12] is advantageously obtained by oxidation of acetyl derivatives of polystyrene (PS).

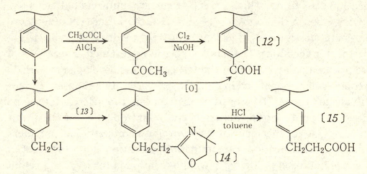

Oxidation of chloromethylated polystyrene (to be explained later) also forms a polymer with carboxylic groups [12]. Chloromethylated polystyrene reacts with 2-oxazolinylmethyl lithium [13], and decomposition of the reaction product [14] with acid produces the carboxylic derivative shown as [15].

Carboxymethyl cellulose (CMC) obtained from cellulose is well known as a water-soluble thickener. Poly(glutamic acid) [16] is the only synthetic polyamino acid whose ester is produced on an industrial scale.

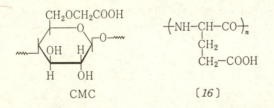

CMC   [16]

The acid anhydride structure is stable enough to store, and shows considerable reactivity. Maleic anhydride (MAN), having low polymerizability, copolymerizes with styrene, vinyl acetate, etc. in the ratio of 1:1, to give a copolymer [17] which contains the structure of succinic acid anhydride in the main chain. On the other hand, addition reactions of MAN to PP, poly(vinyl chloride), PS, and

other polymers having a tertiary CH unit proceed in the presence of a radical
initiator such as peroxide, to produce polymers whose tertiary hydrogen atoms
are partially substituted with anhydrous succinic units [18]. Reactions of MAN

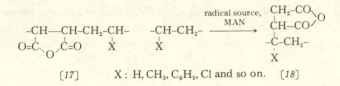

[17]   X: H, CH_3, C_6H_5, Cl and so on.   [18]

with diene-type rubbers like natural rubber proceed more easily. Presumably,
MAN combines with $-CH_2$ in the presence of peroxide [19], while it is introduced
to the double bonds of the main chain during a thermal reaction [20].

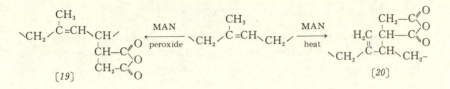

[19]   [20]

These acid anhydride units are useful in activating polymers. When hydrolysed, they become carboxylic acids, and when reacting with alcohols and
amines they give esters and amides. If dyestuffs, pharmaceuticals or enzymes
have amino groups, these reactive polymers can immobilize them. The polymers
with hydroxyl amine [21] are still further dehydrated to give ones containing
units of *N*-hydroxysuccinimide [22]. The polymer [22] reacts with amino
acids to activate them; this is an interesting polymeric reagent in syntheses of

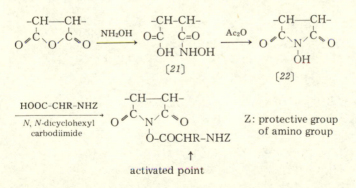

peptides (see §2.3.3 in this chapter). Furthermore an acyl fluoride group is
introduced into the polymer [17] by treatment with sulphur tetrafluoride. Some
acid anhydrides derived from two different acids are known as well. Details of
their synthetic methods will be omitted from this chapter. Here, the main
reaction mechanism for amide formation is summarized: amines attack carbon

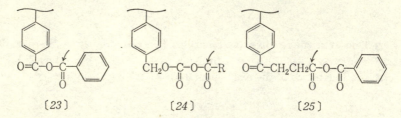

atoms of the acid anhydride structures [23]–[25]. Two C=O groups exhibit electronic and steric effects in promoting the reaction. This is one of the techniques for activation of carboxylic acids which will be explained later.

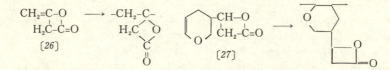

Lactones, as cyclic esters, react with alcohols and amines at high temperature, to give esters and amides. The polymers from ketene dimers [26], and the copolymers from the adduct [27] of acrolein dimer and ketene, are typical lactones with

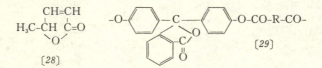

four-membered rings. On the other hand, polymers from angelic lactone [28] and polyester [29] from phenolphthalein (a pH indicator) exhibit reactivity as lactones with five-membered rings. And copolymers containing equimolar

amounts of methacrylate and vinyl chloride form five-membered lactone [30] in the main chain at high temperature.

The reactivity of COOH and COOR groups limits the application of polymers containing such groups to intermediates for reactions performed at high temperature, e.g. reactions with physiologically active substances that are thermally unstable, COOH groups should be converted to more active ones. One method is to convert two different acids into a mixed-type acid anhydride [25] (with activated polymer) but the most simple way is to use an acid halide.

The representative structure of acid chlorides is depicted in [31], derived from acrylic and methacrylic acid (R = H, CH₃). The monomer [31], though troublesome to handle, advantageously gives corresponding polymer [32] by radical polymerization, to which various nucleophiles can be introduced. It is, however, more common in syntheses of various speciality polymers to polymerize monomers [33] obtained by introduction of nucleophiles to [31].

$$
\underset{[32]}{\underset{\overset{|}{\text{COCl}}}{-\text{CR-CH}_2-}} \xleftarrow{\text{polymerization}} \underset{[31]}{\underset{\overset{|}{\text{COCl}}}{\text{CR}=\text{CH}_2}} \xrightarrow{\text{Nu}} \underset{[33]}{\underset{\overset{|}{\text{CO-Nu}}}{\text{CR}=\text{CH}_2}}
$$

The polymer [12] mentioned above gives [34]. The derivative [34] is used for introducing a nucleophilic reagent into the PS structure, but is not used so often as acrylic-type ones.

$$-\text{CH-CH}_2-$$

[34]

Acid chloride is highly active, but unfortunately it is unstable. It is also inconvenient that HCl is produced during the course of reactions. Recent research has been extended to the field of pharmaceuticals and enzymes: pharmaceutically effective units are bound on to the polymer chain, and when the complex is introduced into the body the units are gradually released. On the other hand, immobilization of enzymes makes it possible to perform reactions of enzymes in the solid phase. In these fields, moderate reactivity is required. Activation methods applied to amino acids can also fulfil the requirement, since they involve introduction of *N*-hydroxysuccinimides (e.g. [22]), *p*-nitrophenol esters and amides derived from imidazole and benzotriazole, etc. Activation of polymers has been carefully designed in recent researches, for instance crosslinking of polymers which aim at lowering solubility, or introducing other groups in the vicinity of activated groups for the purpose of controlling hydrophilic or lipophilic properties. Some are illustrated below.

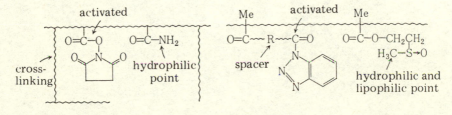

(c) Carbonyl (C=O) groups. Typical polymers exhibiting reactivity due to aldehyde groups are obtained by polymerization of acrolein or methacrolein. The structure of polyacrolein is assumed to be [35], since not only vinyl groups

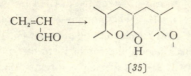

CH₂=CH
|
CHO

[35]

but also CHO groups in acrolein have participated in the polymerization. This acetal structure is, however, easily converted to the CHO structure. For example, treatment of polymers having the acetal structure with NaHSO₄ or NH₂OH gives water-soluble CHO-type precursors (e.g. one with –CH(OH) (SO₃Na)) which can be easily handled. Polymers containing aromatic aldehyde groups can be obtained from chloromethylated polystyrene, as shown below. Polymer [36] is obtained directly; whereas polymer [37] is easily synthesized with an appropriate reagent.

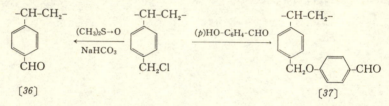

[36]   [37]

On the other hand, numerous acetal-type and epoxide-type precursors have been devised. The former precursors (e.g. [38]) are polymerized by cleavage of the acetal bond to the aldehyde group. Another method to derive a reactive polymer having an aldehyde group is ring-opening polymerization of the epoxy group of [39]. Another polymer with aldehyde groups is [40], which is obtained from starch or cellulose by oxidation with periodic acid. Aldehyde starches, which have already been commercialized, are used as binders for gelatin and other high polymers with amino groups.

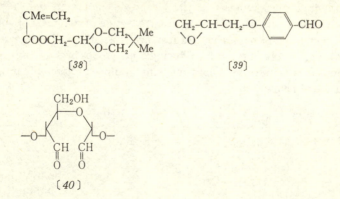

[38]   [39]

[40]

CHO groups are reactive enough to be oxidized or reduced; COOH or CH_2OH groups are obtained. CHO groups react with amines to give Schiff bases. Polymers containing CHO groups are also applied to immobilize enzymes.

Ketones are much less reactive: they are stable enough to be much more easily handled than aldehydes. Typical polymers containing aliphatic or aromatic ketones are easily derived from methylvinyl ketone and acetylated polystyrene, respectively. CO groups are reactive enough to condense with amine to give high molecular nitroxy compounds (e.g. [41], stable radical). A methyl group adjacent to the CO= group is converted to the one like that in [42], which is used as a photosensitive resin.

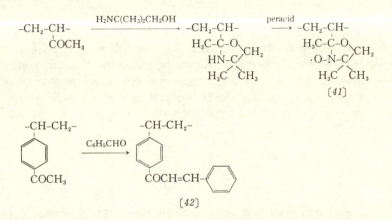

[41]

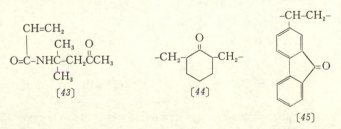

[42]

Other products obtained by partial copolymerization of PE with CO are utilized for sheets for agricultural use. These sheets easily decompose under ultraviolet rays.

Diacetone acrylamide [43] has found practical use; polymers of it tend to become crosslinked when irradiated with light. Resin [44] obtained by condensation of cyclohexanone with formalin, though industrialized as a coating material, should be further studied in order to utilize the reactivity of its C=O groups. The same situation applied in the case of newly developed fluorenone-type polymers [45].

$$CH=CH_2$$
$$\underset{\underset{\underset{CH_3}{|}}{O=C-NH\overset{|}{C}-CH_2\overset{O}{\overset{||}{C}}CH_3}}{\underset{CH_3}{|}}$$

[43]

$$-CH_2-\underset{}{\overset{O}{\overset{||}{\bigcirc}}}-CH_2-$$

[44]

[45]

Polymer [46] is an example containing the alloxane structure, and is a typical polymer with vicinal CO groups. The alloxane-type structure, derived from a barbituric acid-type structure, plays an important role as a reductone *in vitro.* Interesting research projects are being carried out to do with the electron-transfer (redox) ability of the central reactive C=O groups in [46].

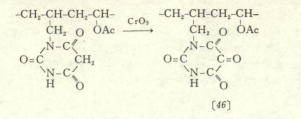

[46]

(d) Ethylenic (C=C), and acetylenic (C≡C) groups. The C=C structure is introduced for the purpose of carrying out various addition reactions, especially crosslinking reactions of polymers. Generally, photoresist-type photosensitive resins should rapidly crosslink and become insoluble as soon as exposed to the light. It is well known that the cinnamic acid ester of PVA crosslinks to cyclobutane in the presence of a photosensitizer when irradiated with light.

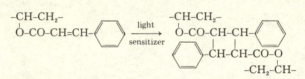

The monomers, which have two or more of the same type of vinyl group per molecule, are polymerized to crosslink and thereby become insoluble under the general polymerization condition. Divinyl benzene (DVB) and dimethacrylate of α, ω-diol [47] are therefore important for crosslinked polymerization, but inappropriate for syntheses of polymers containing pendant C=C groups. Polymerization of monomers containing different reactive groups in the chain is interesting. For example, monomers containing both vinyl and allyl groups polymerize under mild conditions to give vinyl polymers containing unreacted allyl groups, which can participate in subsequent polymerizations or reactions.

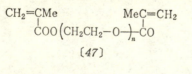

[47]

The differences in properties of groups in a monomer are more remarkable in glycidyl methacrylate [48]. C=C bonds of the acrylic structures in this material are radical or anionically polymerizable, while epoxide structures are cationically polymerized. Therefore, the polymerization in the presence of radical initiator (e.g. azobisisobutylonitrile, AIBN) gives polymers containing epoxide in the side chain; when cationic initiator is used, polymers with vinyl groups are obtained.

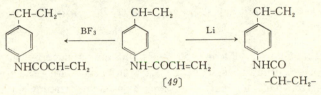

Polymers obtained by radical, cationic or anionic polymerization of monomers [49] are different. The first polymerization gives a gel, the second and the third give polymers which form crosslinks in the presence of radical reagents or under irradiation of light.

$$-CH-CH_2- \quad\quad CH=CH_2 \quad\quad\quad CH=CH_2$$

Monomers containing more than one vinyl group, such as DVB, are polymerized easily to crosslinked polymers, but they can be copolymerized with styrene or MMA monomers in water in the presence of proper emulsifier and water-soluble initiator (e.g. $K_2S_2O_8$ /NaHCO$_3$), to form crosslinked microgels of a diameter around 0.01 m (Fig. 2.1 [50]). Unreacted C=C groups on the surface of the microgels are easily replaced with various functional groups as shown in Fig. 2.1. The microgels can therefore be regarded as reactive polymers. The reaction between residual C=C bonds is carried out by heating; the microgels combine with one another to give insoluble and infusible coating films.

Fig. 2.1. Reactive microgel D is a dyestuff. (From W. Funke *et al.*, *J. Oil Color Chem. Assoc.*, **60**, 438 (1977).)

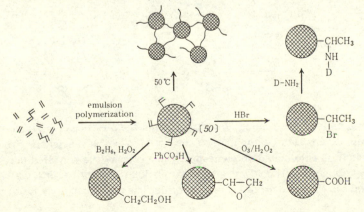

Polyethylenimine containing acrylic groups is advantageously obtained by cationic polymerization of oxazoline [51], since its ring is easily cleaved. The addition condensation of [52] with formalin, on the other hand, gives unsaturated polymers.

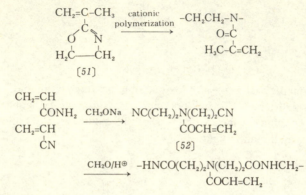

[51]

[52]

Polymers containing vinylidene-type C=C bonds are obtained by cationic polymerization of spiropyranes, like [53]. Little shrinkage during the polymerization is observed. This is an example of expansion polymerization. Esterification

[53]

of OH-containing polymers with acrylic acid and that of COOH-containing polymers with allyl alcohol are examples of practical polymer reactions performed for the purpose of introduction of unsaturated groups. Silica or glass, vinylated in the presence of silane coupler (see § 2.2), is cross-polymerized with unsaturated polyester to a copolymer; this is an important process for manufacturing so-called FRP.

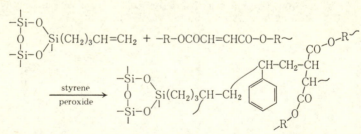

Based on recent technology for vinylation in the presence of vinyl aluminium compounds, unsaturated polymers of the type [54] are advantageously manufactured from chlorinated ethylene–propylene rubber. The vinylated polymers thereby obtained, with a low degree of vinylation (0.5% by weight), easily form crosslinks in the presence of sulphur chloride. The new technology is not only

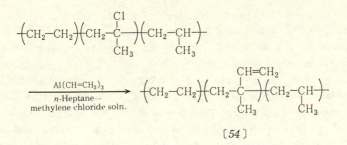

[54]

applied to chlorinated polymers but also to other halogenated ones; these applications are now attracting attention. The preparation of polymers with pendant acetylenic groups – namely polymers of vinyl acetylene – is more complicated than expected. The monomers, whose CH bonds have been protected with trimethyl silyl groups, for example, polymerize under anionic conditions. Thereafter, the protecting groups are replaced with hydrogen atoms to give the desired linear polymers. Monomers which are derived from acrylic acid and contain

$$CH_2=CH \atop \underset{C \equiv C-Si(CH_3)_2}{|} \longrightarrow -CH_2-CH- \atop \underset{C \equiv C-Si(CH_3)_3}{|} \overset{KOH}{\longrightarrow} -CH_2-CH- \atop \underset{C \equiv CH}{|}$$

$C \equiv C$ structures, like [55], have also been studied. However, development of reactive polymers containing the $C \equiv C$ structure is, on the whole, far behind that of polymers with the $C=C$ structure.

$$CH_2=CR \atop \underset{O=C-O(CHR')_nC \equiv CH}{|}$$

[55]

$$-CH_2-CH- \atop \underset{CH=CH_2}{|}$$

[56]

Well-known 1,4- and 1,2-diene-type polymers [56] with $C=C$ structures in the main chain and side chains, respectively, are industrialized as rubber for general use. Those with a low degree of polymerization are typical reactive polymers. Being liquid at room temperature and polymerizable by cleavage of $C=C$ bonds during curing, they are applicable as adhesives, thickeners and sealing materials.

(e) Halogens. Halogen-containing polymers such as poly(vinyl chloride) (PVC), polyvinylidenechloride, polychloroprene and polyepichlorohydrin (PECH) are used as stable industrial plastics. It is difficult to call them reactive polymers. Chlorine contained in the polymers, however, becomes more reactive with various nucleophiles in the presence of certain reagents or under specific conditions. For example, crosslinkage occurs on chlorine in PECH, the reactivity of chlorine in PECH being very important.

  Chloromethylated PS [7] whose high reactivity is attributed to chlorobenzyl-type halogen, forms crosslinks and gives solid granules in the presence of DVB, etc. It is important as a raw material for manufacturing ion-exchange resins, high molecular weight catalysts and reagents.

  Formalin and hydrochloric acid were used for chloromethylation of PS as an intermediate to anion-exchange resin (e.g. $PS-CH_2N^{\oplus}Me_3Cl^{\ominus}$). However, they have been replaced by chloromethylmethylether, which at first proved itself to be a swelling agent for crosslinked PS as well as an effective chloromethylating agent. So-called gel-type granular PS in 200–300 mesh crosslinked with 2% DVB is used for the preparation of polymeric reagents. On the other hand, PS obtained with a third to half of DVB forms a highly porous aggregation, called MR (macroreticular) resin or high-porous type PS, and does not undergo volume change when suspended in solvents.

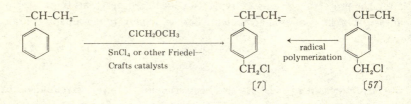

[7]   [57]

  Though less satisfactory in mechanical strength, chloromethylated PS granules obtained from MR-type PS are more widely used than gel-type PS in reactions that proceed in organic solvents. This is because they never collapse even when inner strain becomes high with swelling of the granules.

  Furthermore, [7] can be manufactured by polymerization of chloromethyl-styrene [57]; copolymerization of [57] with DVB will lead to [7] crosslinked with DVB. Compound [57] is now commercially available as a result of successful chlorination of vinyl toluene at high temperature, which is more advantageous than the conventional troublesome method. Commercial chloro-

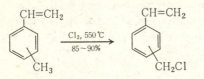

methylstyrene is now a mixture of 40% *m*-isomer and 60% *p*-isomer; this is acceptable for general use. Conventional methods for the synthesis of the pure *p*-isomer are complicated. Various synthetic methods have been devised, and an example is cited below. This method, though far from industrialization, is promising because it employs no chloromethylether, which is reported to have carcinogenic properties. Compound [7] exhibits reactivity analogous to that of

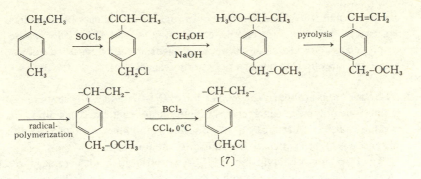

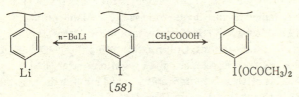

[7]

benzyl chloride; its utility is discussed with several examples in §2.2.2 and other places of this chapter. Chloromethylstyrene [57] is a typical M—R (see p. 5), prevalently converted to M—F and P—F (functional monomer and polymer, respectively) and applied in various fields. When the reactivity of CH_2Cl groups is not high enough for a certain reaction, potassium iodide is advantageously added into the reaction system. Then CH_2I groups are formed, which enable the reaction to proceed. To convert CH_2Cl groups into CH_2F groups is also possible; however, the reactivity of the latter groups has not yet been clearly examined.

In contrast, the reactivity of halogen directly attached to the aromatic ring is remarkably low. Styrene with a chlorinated benzene ring has been commercialized on a small scale; the chlorine of these polymers has little reactivity under normal conditions. Exceptionally, the reactivity of iodinated PS [58] obtained by treatment of PS with I_2/P_2O_5 is quite high and can give a lithium salt or iodoso derivative.

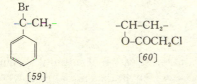

[58]

For other reactive halogen derivatives, α-bromo PS [59] and chloroacetates of PVA and cellulose (e.g. [60]) are not utilized to the same extent as [7].

Br
-C-CH₂-   -CH-CH₂-
                 O-COCH₂Cl
                 [60]

[59]

For inorganic, chlorinated high polymers of silica, refer to §2.2 of this chapter. Halogens attached to N, O and other hetero- atoms exhibiting unique reactivity, and *N*-halogens in amide or imide, show high cationic or radical reactivity. Therefore, for example, chlorinated nylon (-CO—NCl-) can be employed as an oxidizing or chlorinating agent; it can be regarded not as a reactive intermediate but as a polymeric reagent.

(f) Nitrogenous groups (-NH₂, NH, N—CH₂OH, etc.). Development of amine-type polymers, which contain highly important reactive amino groups, is somewhat behind that of COOH-containing polymers, owing to the difficulty in synthesis and the intrinsic instability of the polymers.

The instability of vinyl amine ($CH_2=CH-NH_2$) makes it necessary to protect $-NH_2$ groups in monomers during polymerization. The methods most often proposed involve protection of $-NH_2$ groups in the form of amide or urethane prior to polymerization, and subsequent regeneration of the $-NH_2$ groups. Poly(vinyl amine hydrochloride) [61] is synthesized with low-cost raw materials and in high yields by the recently developed method, given below.

$$CH_3CHO \atop CH_3CONH_2 \xrightarrow{H_2SO_4} CH_3CH {\nearrow NHCOCH_3 \atop \searrow NHCOCH_3} \xrightarrow[\text{celite}]{180°C} {CH=CH_2 \atop NHCOCH_3}$$

$$\xrightarrow{AIBN} {-CH-CH_2- \atop NHCOCH_3} \xrightarrow{HCl} {-CH-CH_2- \atop NH_2 \cdot HCl}$$

[*61*]

A question that always arises is by which method can the desired amine polymers be most advantageously obtained – by polymerization of the corresponding monomers or by reaction of the polymers proper? Polymers of the type [61] are also obtained by heating commercially available polyacrylamide in the presence of bromine–alkali (Hoffman reaction). However, the method involves many side-reactions. Chlorination using Cl_2/NaOH at around 0 °C, which has recently been developed, gives a white powder [61] in a yield of more than 90%. This method is remarkably simple and practical; however, the COOH content in this polymer is less than 10%.

$$-CH-CH_2- \atop CONH_2 \xrightarrow[\text{2)HCl}]{\text{1)Cl}_2/\text{NaOH}} {-CH-CH_2- \atop NH_2 \cdot HCl}$$

[*61*]

For aromatic amine-type polymers, polymerization of substituted aminostyrene is usually performed; this process is simpler than nitration of PS followed by reduction. Reduction of the starting monomers with phenylhydrazine gives virtually pure polymers. Also, the Schmidt reaction process of acetylated PS gives the corresponding reaction products in high yield.

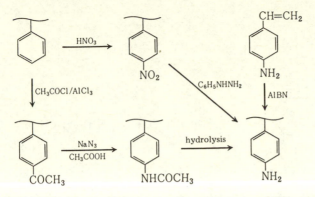

In order to make good use of the original reactivity of NH_2 groups, it is advisable to separate them as far as possible away from the main chain. Benzyl amine-type polymers, like [63], can be obtained by treatment of chloromethylated polystyrene with phthalimide (Gabriel reaction) or with liquid ammonia or reductive amination of formylated PS.

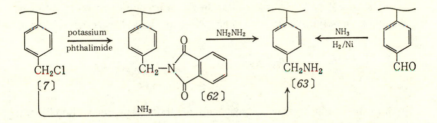

Though the reaction itself is advantageous after the Gabriel reaction is carried out, CH_2Cl groups remain unreacted. *N*-chloromethylphthalimide is prepared to react with PS. Decomposition of the reaction product with hydrazine gives [63] which is free from unreacted CH_2Cl groups; but the amount of methylphthalimide in PS is relatively low.

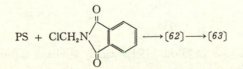

There are many polymers which exhibit high reactivities as amines. Into these polymers are introduced various functional groups as follows: polymer [64] is obtained by etherification of cellulose with nitrobenzyl chloride followed by reduction, polymer [65] by treatment of glass beads with silane coupler, chitosan [66] by hydrolysis of chitin, and poly(amino acid) [67] by polymerization of

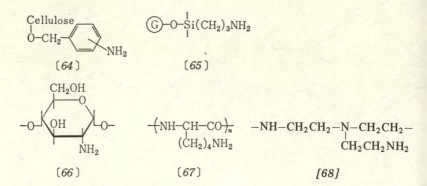

[64]　　　　　　　[65]

[66]　　　　　　　[67]　　　　　　　[68]

lysine. Those NH_2 groups combined to triazine (e.g. in melamine, or guanamine) and to pyrimidine rings (e.g. in nucleic acids) show weak basicity, so that they are not suitable as reactive intermediates.

A typical polymer containing secondary amine is polyethylenimine. The commercially available polymer is derived from ethylenimine and contains many branches; there are many tertiary and primary amines in its branches [68]. Polyethylenimine beads crosslinked by epichlorohydrine have recently found practical use. Polyethylenimine containing no branches can be obtained by ring-opening polymerization of oxazolidone or 2-alkyloxazoline followed by

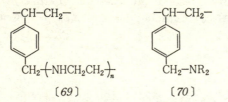

hydrolysis. For the purpose of scavenging metallic ions, the secondary amine is combined with PS as in [69], which has already been practically used. Tertiary amine-containing polymers have been developed, too: for example polymer [70] is used as an ion-exchange resin of weak basicity for special cases. The acrylic [71]

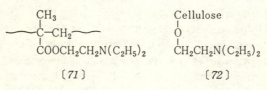

[69]　　　　　　　[70]

type is also well known, and DEAE cellulose [72] is widely used as a stationary phase in liquid chromatography.

CH₃
|
——C—CH₂——
|
COOCH₂CH₂N(C₂H₅)₂

[71]

Cellulose
|
O
|
CH₂CH₂N(C₂H₅)₂

[72]

(g) Epoxides. The reactivity of epoxides is well known. A so-called epoxide resin
is a low molecular weight substance (or oligomer) having epoxide structures at
both ends of the molecule. When treated with a curing agent, the monomer (or
oligomers) cleave the epoxide rings so as to leave –OH groups in a three-dimensional
structure, and accordingly show an adhesive property. An epoxide resin therefore
cannot easily be classified as a reactive polymer.

As mentioned above, vinyl polymers derived from glycidyl methacrylate [48]
react with various nucleophiles and electrophiles at their pendant epoxide groups.
Porous glass has been treated with [48] and thereby epoxidated on the surface
for the purpose of introducing various functional groups. Polypropylene (PP)
or acrylic fibre with small amounts of [48] combine with dyestuffs containing
amino groups, which are called reactive fibres.

$$
\begin{array}{c}
CH_3 \\
-\overset{|}{C}-CH_2- \\
\overset{|}{C}OO-CH_2-CH-CH_2 \\
\diagdown O \diagup
\end{array}
\quad + \; H_2N\text{–dye} \quad \longrightarrow \quad
\begin{array}{c}
CH_3 \\
-\overset{|}{C}-CH_2- \\
\overset{|}{C}OOCH_2CH-CH_2-NH\text{–dye} \\
\overset{|}{O}H
\end{array}
$$

Among conventional polymers, those containing C=C bonds, like diene-type
rubber, can be epoxidized with peroxy acid. For example poly(1,2-butadiene) [56],
which contains C=C bonds in its branches, is more easily epoxidized than other
polymers which contain the double bonds in the main chain. The polymer [73] is
in practical use as a reactive oligomer. Also, OH containing polymers, like cellulose,

$$
-CH_2-CH=CH-CH_2- \xrightarrow{\text{peracid}} -CH_2-CH-CH-CH_2-
\underset{\diagdown O \diagup}{}
\qquad
\begin{array}{c}
-CH_2-CH- \\
\overset{|}{C}H-CH_2 \\
\diagdown O \diagup
\end{array}
$$

[73]

are esterified with unsaturated acid, then oxidized to give epoxide. By etherification of phenol-formalin resin with epichlorohydrin, an epoxide such as [74]
is obtained.

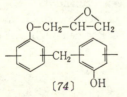

[74]

Polymerization (hardening) of epoxy resins, which are not reactive polymers,
can be promoted by introduction of functional groups. There are many studies
concerning introduction of crown ether into polymers. For example, when epoxy
prepolymers are cured in the presence of crown ether containing amino groups,
which acts as a hardener, the crown ether combines with the epoxy resins. It is
a very simple and advantageous method.

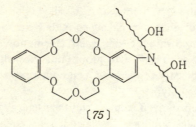

[75]

(h) Other reactive groups. Similarly, there are many kinds of compounds that can be regarded as reactive polymers; reactive functional groups are many in the field of organic chemistry.

It is possible to obtain an isocyanate structure (NCO) by polymerization of vinyl isocyanate or $CH_2=CH-C_6H_4-NCO$. However, it would be extremely difficult to synthesize such monomers and then polymerize them because of their remarkably high activity. PS-NCO is obtained by reaction of PS-NH_2 with phosgene. In this case, however, the reaction is not always complete nor desirable, since a reaction occurs between the formed -NCO and the remaining -NH_2. An excellent way to obtain the isocyanate structure is by addition of NCO derivatives shown by [76] to C=C in rubber.

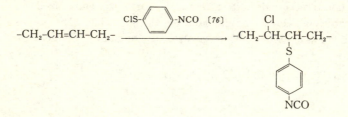

Several kinds of diisocyanates have been in practical use in the production of polyurethane. An appreciable difference has been found in the reactivities of both NCO groups when the compound has an asymmetrical structure, like 2,4-toluenediisocyanate (TDI). In a reaction between TDI and the -OH groups found in a modified polymer, e.g. PE - which was partially oxidized, then reduced - the polymer [77] was derived, since 4-NCO, which is more reactive

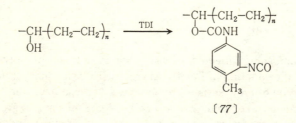

[77]

than 2-NCO, reacted with –OH groups, but 2-NCO did not. A block polymer [78] was also obtained by anionic polymerization of 4-NCO of TDI at the anionic terminal of a living polymer.

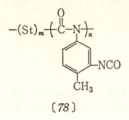

[78]

By reaction of NCO groups in the presence of a trace amount of H_2O, which exists in the system, crosslinking or other reactions easily occur because of the high activity of the NCO group. Therefore, to avoid such unfavourable reactions, the reactive group should be appropriately preserved until the desired reaction is performed. The use of precursors is very important in designing highly reactive polymers. Thus such compounds, so-called amine–imide-type ones, have been designed. For example, N-ylide shown by [79] is one of the very useful NCO precursors, because it decomposes at temperatures higher than 130 °C to give

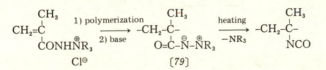

[79]

–NCO. Compound [80], which is a copolymer with styrene, was developed for similar reasons. In this case, the NCO groups formed are found to be stable

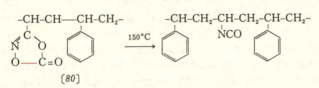

[80]

against moisture even at relatively high temperature because they are surrounded by hydrophobic styrene structures. The structure of $\supset N$–CH_2OH formed from the reaction between amine and formalin is remarkably unstable and it easily forms $\supset N$–CH_2–$N\subset$. However, when it combines with amide the OH group is stabilized by the formation of a hydrogen bond with $\supset C=O$. Then it exhibits very strong electrophilic properties through the formation of reactive species such as $\supset N$–$CH_2^{\oplus} \leftrightarrow N^{\oplus}=CH_2$ in the presence of acid catalysts. This process is found during the formation of urea resin or melamin–formalin resin, and can be applied in the activation of amide-type polymers. The polymer [81] obtained

from *N*-methylol acrylamide easily crosslinks in the presence of weak acid to
produce materials which are called thermosetting acrylic resins. It is also applied
when one wants to introduce nucleophilic agents (Nu: amines, phenols) into
polymers as shown by the following reaction. *N*-methylol acrylamide is obtained

$$-CH-CH_2- \xrightarrow{\ Nu\ } -CH-CH_2-$$
$$\underset{\ }{CONHCH_2OH} \qquad \underset{\ }{CONH-CH_2Nu}$$
[81]

as a monomer, and is used in the crosslinking or modification of cellulose, since
its vinyl or methylol groups independently react with cellulose. *N*-methylol

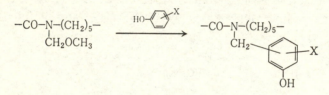

compounds of Nylon are usually synthesized as methylethers and it becomes
possible to introduce nucleophilic agents in the same way.

$$-CO-N-(CH_2)_5- \xrightarrow{\ \text{HO}-\langle\bigcirc\rangle^X\ } -CO-N-(CH_2)_5-$$

The isothiocyanate group (NCS) is preferable in practice to the NCO
group, since the former is less active than the latter. The NCS group easily reacts
with NH_2 groups but not with H_2O or COOH groups. The NCS group fixed in
the polymer reacts with terminal NH_2 groups in peptides whose configurations
are undefined. As shown by the following reaction, the new peptide is eliminated
by treatment with a strong acid. Loss of the terminal amino acid from the
original peptide simultaneously occurs, together with cyclization to form
a thiohydantoin. This is a variation of the Edman degradation method, which
is one of the techniques to determine the sequence of amino acids in peptides.

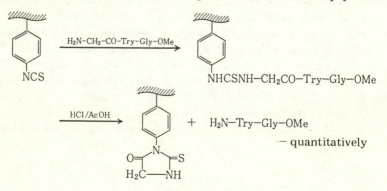

The thiol group (SH) has a high reactivity and a specific property: it is characterized by its reducing power and by the formation of mercaptide with metals. The SH group is also interesting in relation to the cysteine-containing peptides glutathione, co-enzyme A and lipoic acid, which play important roles in living organisms. $CH_2=CH-SH$ is very unstable, like vinylamine. In the synthesis of polyvinylthiol from the corresponding monomer, it is necessary to protect SH groups because they retard radical polymerization. Polyvinylthiol has been mainly synthesized by hydrolysis of the polymer obtained from the radical polymerization of vinylthioacetate. An alternative method for the preparation of polyvinylthiol has also been found, and is shown by the following pathways.

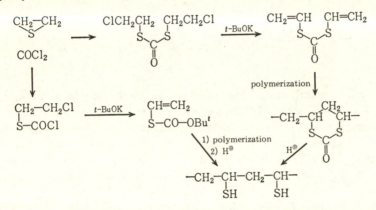

Other methods using condensation polymerization are available, and a typical example is shown below. In general, oxidation of SH groups is extremely easy,

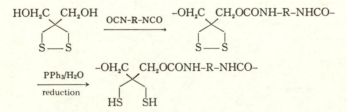

and then gelation (crosslinking, insolubilization) of the polymers which contain SH groups easily occurs as a result of the formation of intermolecular S-S bonds. Therefore, a small number of SH groups are usually introduced into the polymer and the compounds [82], [83] and [84] are synthesized from the corresponding polymers.

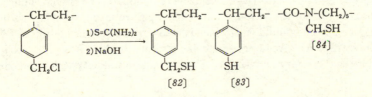

A compound with an S–S bond is used as a precursor to introduce SH groups into the polymer. For example, lipoic acid [85] is used as follows: lipoic acid reacts with an NH_2-containing polymer (shown in [63]) after its activation; or the compound [86], which was obtained by monoamidation of lipoic acid with diamine, reacts with BrCN-activated cellulose on COOH-activated polymers. Dihydro-type polymer is obtained by subsequent reduction. The oxidation–

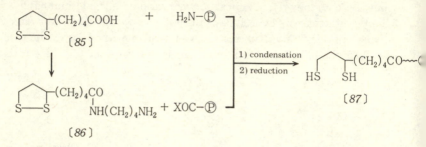

reduction potential between lipoic acid and dihydrolipoic acid, E (pH = 7), is 0.32 V; the compound [87] thus possesses a strong reducing power. Therefore, it is possible to reactivate an oxidized or deactivated (as –S—S–) enzyme (for example, papain, which contains the cysteine structure) by using [87] as a reducing agent. The polymer [87] itself is oxidized to lipoic acid type in the reaction. It is possible to use [87] repeatedly after its reduction to the original form.

The formation of C–MgX (Grignard reagents) structure in a polymer is not easy. Nevertheless, corresponding C–Li structures are found in several cases, especially in PS; this reactive polymer is widely used. The compound [88] is obtained from reaction between bromo-PS or iodo-PS (crosslinked and granular type) and *n*-butyl lithium. It reacts with fluorene to produce the corresponding lithium compound.

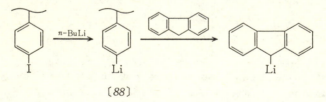

[88]

More useful applications are also found in reactions with electrophilic reagents in which various functional groups are introduced; typical examples are shown in Fig. 2.2.

Reactive groups – azide (N_3), acid azide (CON_3), etc – are stable enough, and various methods for the introduction of those groups into polymers have been developed. The compound [89] which is easily obtained by azidation of PVC, is an interesting reactive intermediate like phosphiniamine (P–N-ylide [90]) which is the secondary derivative of PVC. Aromatic amine-type compounds

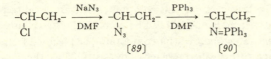

[89]   [90]

are easily diazotized and the products generally give either azo-type dyestuffs by reaction with naphthols without separation, immobilized enzymes by coupling a phenol group in the enzyme (e.g. in tyrosine), or azide compounds by reaction with sodium azide.

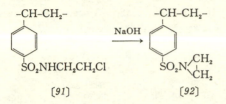

[91]   [92]

More skilful techniques are required for the reaction using an aziridine ring, since it is more reactive than epoxide. The compound [91] can safely be stored and is available as a precursor that can easily produce the aziridine structure [92] by reaction with alkali.

In considering reactive polymers, it is very important to know how to use multifunctional, low molecular weight compounds such as silane coupler. Cyanuryl chloride is a highly reactive halogenated compound that is practically

Fig. 2.2. Reactions of lithium polystyrene.

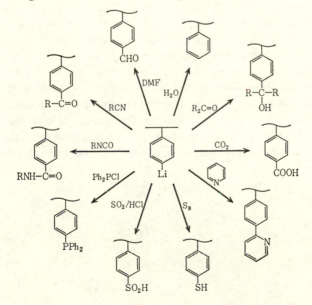

used. However, it is difficult to preserve this compound when bonded to polymers. Since an amine-type dyestuff molecule (D) can be combined with cellulose, according to the reactions shown below, the compound [93] formed from cyanuryl chloride and dyestuff is called a reactive dyestuff. Thus, the dyestuff is fixed by reaction between [93] and cellulose. On the other hand, the compound [94], which is obtained from a condensation reaction between cellulose and cyanuryl chloride, is called a reactive fibre, and it can fix dyestuff [95] by reacting with it. The reactive dyestuff is in wide use: it can be safely preserved and the required amount can be conveniently prepared by treatment with the appropriate amount of dyestuff. The reactive fibre, however, has not been commonly used because it has many disadvantages compared with the reactive dyestuff.

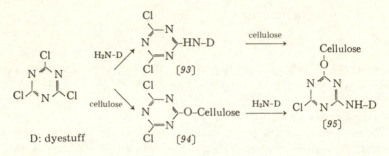

D: dyestuff

Relationships between reactive polymers (reactive intermediates) and various reactive groups have been described above. It should be emphasized that the type of carrier is also important for selecting reactive groups. For example, silica or alumina particles are coated with polyethylenimine (PEI) film by immersing them in an aqueous PEI solution. Thus, the surface of the particle is covered with reactive -NH- groups, but that film is easily removed by washing with water. By treatment with glutaraldehyde, imino groups on the surface PEI are partially crosslinked. Thus, the film can be fixed at a solid surface and it becomes insoluble in water. This is a suitable method to obtain very desirable reactive solid particles. The solid core can be removed by treatment with strong acids. The reactive PEI which possesses a very wide surface area can be obtained because of the dissolution of the alumina core. In fact, enhanced catalytic activity is obtained by bonding of imidazole to the PEI obtained in that way.

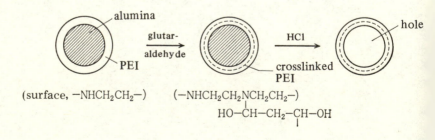

Granular reactive polymer is preferable when it is used as a reagent or a packing agent in column reactors. Such polymers are also available as thin films when one wishes to use them in a bioreactor. There are other proposals for the application of such reactive polymers: e.g. coating materials for electrodes, hollow fibres, artificial organs or electronic elements. Beside such small-scale applications, there is demand for some large-scale applications, e.g. surface coatings for construction materials or the hulls of ships.

2.2.2 Classification of polymers by structure

The reactive groups described in the preceding sections are supported by polymers of various structures. The intrinsic reactivity of polymers used as raw materials and the properties of the carriers are important in determining

Fig. 2.3. Chemical reactions of PE and PP.

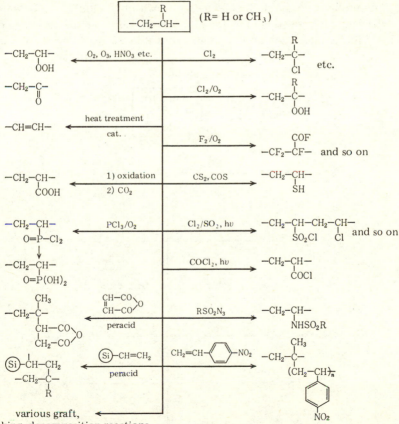

the reactivity of the product reactive polymers, and even more so when the desired polymers are prepared by further reaction. In the preceding section we discussed with many examples the relationship between the structure of polymers and the method of introducing reactive groups; we now discuss conventional polymers. PE and PP cited in Fig. 2.3 are remarkably stable; they are representative of polymers that hardly react ionically.

The number of reactive groups introduced in PE or PP is very low. However, chemical modification itself is an effective way to alter the stable hydrophobic structure of PP or PE. We should note that PP, the cheapest industrial plastic material, can now be endowed (though on a trial basis at present) with very refined functions, such as anchoring of enzymes. Reactions involving PE and PP, which are used on an industrial scale, need to be applied in practical, industrial processes like moulding, spinning and film-making, rather than in laboratory work. Copolymerization of PP or PE with reactive or functional monomers and their graft polymerizations are worth explaining.

Reactions of PS are shown in Fig. 2.4. PS with hydrophilic properties is obtained by introduction of hydrophilic groups. However, the advantage of PS lies in its hardness: here we employ PS granules in the reactions; these granules

Fig. 2.4. Chemical reactions of PS.

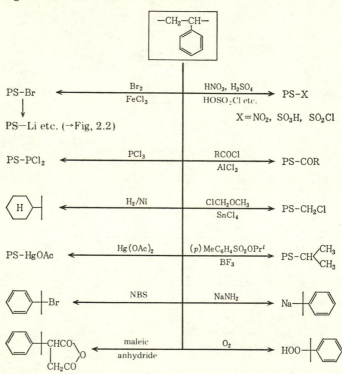

are rendered insoluble by crosslinking in the presence of DVB. Ion-exchange resins are derived from the PS granules. PS is insoluble in water, but becomes soluble when SO_3H groups or other hydrophilic groups are introduced. When crosslinked polymers are used, all reactions proceed in heterogeneous systems. The size of granules and the size and number of pores in granules depend on the quantity of reactive groups introduced. In order to have a high ion-exchange capacity, only one SO_3H or $N^{\oplus}R_3X^{\ominus}$ is introduced per phenyl group in the polymer, but the reactivity of the introduced group is affected by neighbouring groups. The number of reactive groups introduced is generally controlled at around 10%, when the obtained polymer is used as a carrier for peptide solid phase synthesis, or as a carrier for enzymes or transition metals (in catalysts). Generally, the reactivity of a crosslinked polymer (in a heterogeneous system) is lower than that of a soluble one (in a homogeneous system). Reactions of PS-Li were shown in Fig. 2.2; some reactions of the widely used chloromethyl-ated PS, are shown in Fig. 2.5.

Reactions of cellulose (a natural polymer) have been investigated for many years. Cellulose is itself a reactive OH-type polymer. The original hydrogen

Fig. 2.5. Chemical reactions of PS–CH_2Cl.

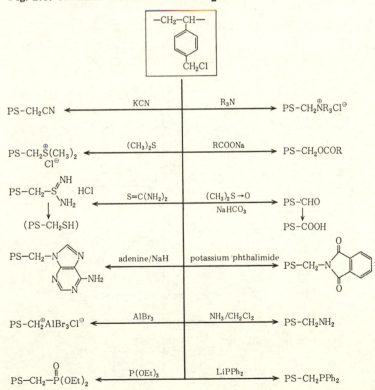

bonds in cellulose are cleaved by esterification and etherification: in this way its solubility and affinity for organic solvents are enhanced. The reactivity of cellulose is further increased because each reactive group exists far from the main chain. Especially, the introduction of COOH or NH_2 groups into cellulose is a method to produce a novel cellulose derivative with functional groups. Some examples of reactive cellulose are shown, without giving methods of preparation, in Fig. 2.6. Cellulose should not be considered as an unpromising material, because it is an inexpensive, renewable, reactive and versatile material.

Reactions of silica and glass, typical inorganic high polymers, are demonstrated in Fig. 2.7. Since it is chemically stable, glass has been conventionally used only as containers for reactive chemicals, and the reactivity of glass has not been studied in detail, except with regard to its dissolution in hydrofluoric acid. The -Si—O—Si—O- type siloxane structure in glass contains OH groups at the ends of the main chain. Crosslinked, or three-dimensional silica contains numerous OH groups, and the high adsorbing ability of porous silica gels is proved when they are used as a packing material for liquid chromatography. However, when they exist in high density, the OH groups form hydrogen bonds and cease to exhibit their original reactivity, so that cellulose can not be considered as an acidic OH-containing polymer. Treatments to enhance the reactivity

Fig. 2.6. Reactive cellulose.

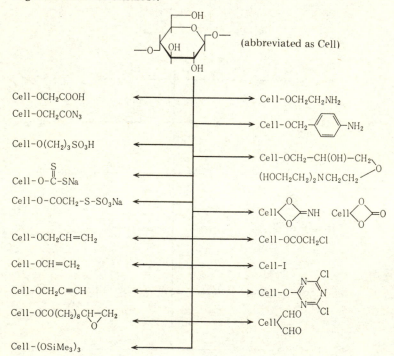

of silica and glass have become simple, especially since silane coupler was developed. The silane coupler is generally represented by $X_{4-n}Si-R-Y_n$: X is a functional group such as RO or Cl which can be hydrolysed. Y is $CH=CH_2$, NH_2, Cl, COOH, SH, epoxide or some other functional groups to be introduced into the polymer. In the treatment of silica with the silane coupler, X reacts with

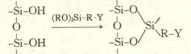

$\geqslant$Si—OH at the surface of the silica and the surface is functionalized with Y, as shown above. In such a way, the surface of silica and glass is activated. By using the reactivity of Y, various reactive groups are introduced onto silica and glass. Examples are shown in Fig. 2.8.

Fig. 2.7. Reactions of silica (glass).

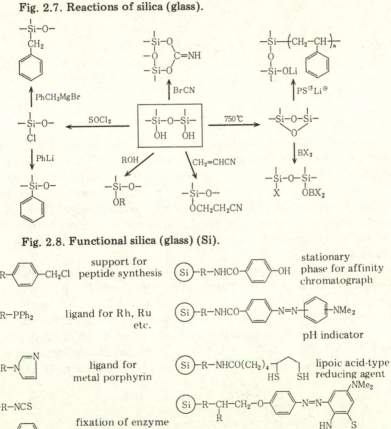

Fig. 2.8. Functional silica (glass) (Si).

What is more important is that the technology of silane coupler has been extended from silica to the majority of inorganic materials involving alumina, tin oxide and ferrite. The silane coupler has recently been used to modify chemically platinum and gold electrodes. Furthermore, reactions and activations of natural resources which could not be effectively used are now intensively examined; carbon materials such as activated charcoal, carbon black, and coals, polysaccharides such as chitin, humin, and lignin, and structural proteins such as collagen and keratin have not yet been utilized widely. The following example shows an attempt to give asymmetric selectivity to electrolytic oxidation-reduction reactions by introduction of asymmetric amino acids [96] onto an activated COOH structure on the surface of a carbon electrode.

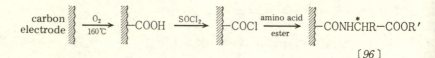

[96]

2.3 Application of reactive polymers

As we have seen, various reactive groups can be introduced into parent polymers. In this section, the application of the reactive polymer thus obtained is discussed briefly.

2.3.1 *Reactive polymers for graft and block polymerization*

Various attempts have been made to develop new materials by the binding of a second polymer chain to ready-made polymer (graft polymerization). However, only one method is used in practice: that is polymerization of monomer (M) in the presence of a parent polymer (chain transfer method). Other methods could not be used, because they were expensive. The chain transfer method is indeed simple, but the efficiency for grafting is generally low, and a large number of ungrafted polymers (homopolymer of M) are present in this system.

A more effective method is to use reactive groups in the parent polymers. Namely, an appropriate reactive group is introduced into the parent polymer and then the polymerization is allowed to proceed at the reactive group. The compound [97], which possesses a reactive peroxide group, will be grafted from a radical formed at the polymer by a M-M-M... type branch chain. In this case, the formation of a M-M-M... type homopolymer simultaneously occurs with a low molecular weight initiation (PhCOO·, Ph·). The surface of PE or PP can be

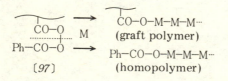

[97]

oxidized to produce a small amount of hydroperoxide [98]. Then, graft polymerization is initiated by the decomposition of this hydroperoxide. This redox
initiation is more effective than radical graft polymerization, since OH· formation
would be suppressed. By extending this principle, ionically polymerizable

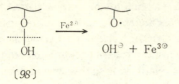

[*98*]

monomers are polymerized from the cationic or anionic centre formed at the
main polymer chain. An aluminium derivative [99] can be derived from vinyl

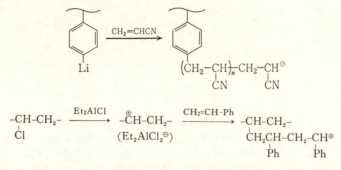

groups in the branch chain, and grafting is possible in the presence of titanium
trichloride through Ziegler-type coordinated anionic polymerization. PVC containing a dithiocarbamate structure [100] can be cleaved at the position indicated
by the dashed line, and the PVC–S formed can initiate the graft polymerization
of co-existing monomers. The parent polymer [101] which possesses an epoxide
group is grafted nucleophilically by PS living anions. Thus, the application of
reactive polymers for graft polymerization is extremely wide. Cationic ring-

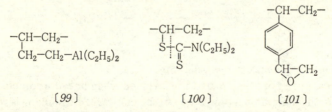

[*99*]   [*100*]   [*101*]

opening block polymerization of oxazoline is a preferable method for introducing many –NH– groups into polymers.

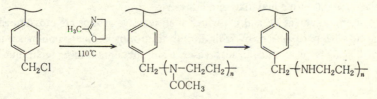

Block polymers are obtained by polymerization of the second monomer onto the reactive group at the end of the polymer. The terminal reactive group is prepared by using an initiator of special structure or by condensation polymerization in which an excess of one component compared with another is used. When a highly active species having an anionic or cationic part at the terminal is formed *in situ*, living polymers are obtained. We can consider many variations when reactive groups such as COOH, NH_2, and SH are at the terminal. In the case of oligomers of silica or dienes, the role of the terminal group is important, since its molecular weight is relatively low; this type of reactive polymer is frequently used for block polymerization or for increase in molecular weight by coupling of terminal groups. The methods used to obtain reactive

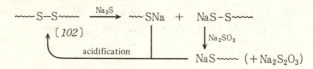

oligomers with reactive terminals by cleavage of the main chain are also interesting. For example, a terminal $\supset$C=O group is obtained by a cleavage of -CH=CH- in the main chain of diene rubber with O_3–NH_2NH_2; a terminal -CHO group is formed by cleavage of -CH(OH)—CH(OH)- (head-to-head bonding) in poly(vinyl alcohol) with HIO_4. Thiokol-type sealant (disulphide-type polymer [102]) is given fluidity by reductive cleavage with sodium sulphide. It is then recombined and solidified by oxidation or bond formation through a salt bridge of the terminal -SH group.

$$\text{~~~S–S~~~} \xrightarrow{\text{Na}_2\text{S}} \text{~~~SNa} \;+\; \text{NaS–S~~~}$$

2.3.2 *Reactive polymers for crosslinking*

Crosslinking or gelation are to be avoided when linear or soluble polymers are the desired final products. On the contrary, crosslinking is remarkably important in order to obtain toughness in cellulose or rubber, stability towards water in ion-exchange resins, or insolubility and stability towards solvents in colouring materials and adhesives.

Polymers such as PE and PP, which lack reactive groups are crosslinked by heating with peroxides or by radiation. C=C groups in diene rubbers are used for crosslinking with sulphur. In such cases, a suitable crosslinking technique is chosen to suit the low activity of the relatively stable polymer. Many attempts are made to introduce suitable reactive groups in order to facilitate crosslinking.

A polymer into which a small amount of a reactive group, like COOH, is introduced can be easily crosslinked by intermolecular salt formation; it can also be easily crosslinked by polyamine, diol, bis-epoxide and diisocyanate. Ethylene–propylene rubber possesses good physical properties, but it cannot be crosslinked because it has no C=C structure. It can be crosslinked after introducing a small amount of an unconjugated diene such as [103] or [104]. Vinyl or allyl groups in main or side chains possess a characteristic stability, and those groups can be easily polymerized by free-radical methods using vinyl monomer or bis-acrylate under suitable conditions.

$CH_3-CH=$ ⬡

[103]

$CH_3-CH=CH-CH_2-CH=CH_2$

[104]

A large number of crosslinking reactions can be designed by using the reactivity of various reactive groups, as described in §2.1 of this chapter. However, only some of them are practical, for there remain many problems of stability, reactivity, difficulty for synthesis, crosslinking density and solubility among the different types of polymer.

Styrene–butadiene block polymer apparently behaves as if it were crosslinked at room temperature because of cohesion between PS segments, but it is fluidized by heating because cohesion is thereby weakened (reversible crosslinking). Thus, the styrene–butadiene block polymer is expected to become a useful material. The following polyester crosslinks by the Diels–Alder type reaction by heating at 100 °C. Dissociation of Diels–Alder-type bonds (reversible Diels–Alder reaction) occurs by heating; hence the crosslinking is reversible.

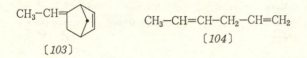

The crosslinking points themselves should be thermostable in heat-resistant polymers. Polymers which include the biphenylene structure are studied in relation to synthesis of polyimides and polyesters. Such polymers form very strong and irreversible crosslinks at temperatures higher than 300 °C. The crosslinking may occur through cleavages of C—C bonds and intermolecular recombination of the formed species as shown by the following reactions:

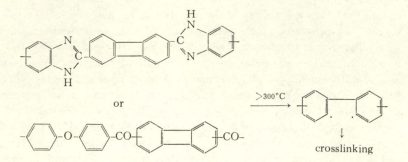

or

crosslinking

Photosetting-type photosensitive resins are obtained when crosslinking is rapidly caused by light. Films of PVA-cinnamate [105] or azido derivatives [106] rapidly crosslink and become insoluble when exposed to ultraviolet light in the presence of a photosensitizer. The non-exposed parts are removed by dissolving with appropriate solvent.

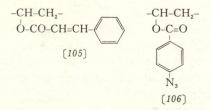

2.3.3 Speciality polymers based on reactivity

Several speciality polymers are intentionally designed to function on the basis of chemical interaction. The most typical example is ion-exchange resins, and methods for their synthesis and use are well established. Resins more skilfully and finely designed are the chelate (formation) resins. There will be further progress in this field because utilization of metal chelate properties (crosslinking by metal ions, electric conductivity, chelate colouring, catalytic activity, bio-activity, etc.) is expanding, and also because the resins have a variety of structure.

Oxidation–reduction resins are characterized by electron exchange (transfer) accompanied by the exchange of metal ions. Much attention has been focussed on the oxidation–reduction reactions because of their importance and great potential for future application. Extensive studies have been done but practical applications are not yet established. Reactive polymers as oxidants or reductants are mentioned later.

Photosensitive resins appeared after ion-exchange resins. A variety of physical properties of these substances are affected by irradiation with light, e.g. photoelectric conductivity, photoelectromotive force, and photosensitive colour development. In general, photosensitive resins are characterized by their photo-resistance property which is due to the crosslinking or insolubilization caused

by light irradiation. An example is the cinnamic ester of PVA, or azido derivatives. The utilization of photodepolymerization of crosslinked materials is
promising. Recently, the technique of having a detailed image on the polymer
film directly (without negative film) has been developed; X-rays or electron
beams are used which have higher energies than visible light. In fact, this technique has been used in super-LSI production. Photodecomposition occurs in
polymers having a ketone structure in the main or side chains. In the case of
electron beams, decomposition of vinylidene-type polymers can be utilized.
Compound [107] decomposes by electron-beam irradiation, generating
isobutylene and SO_2.

$$-(-CH_2-CH-)-SO_2-$$
$$\underset{\displaystyle \underset{CH_3}{|}}{\overset{\displaystyle |}{CH_2}}$$

[107]

The study of polymers developed with the purpose of producing more
stable and durable materials. Recently, the idea of decomposable polymers has
appeared; it originated from the 'vinyl' films used in agriculture. Desirable
polymers for 'vinyl' cloches should possess the following properties: during the
early growing period of plants, the films should be strong enough to keep the
inside temperature warm, and should permit sunshine to pass through; after
the plants grow up, the film should disappear by photodecomposition, thus
without incurring labour. A copolymer of ethylene and CO was developed for
this purpose. However, when the photodecomposable polymers are covered
with soil, further decomposition of the polymer is not possible because the
light does not reach the polymer. Therefore, it is necessary to invent a film that
will decompose by hydrolysis or by other reactions of soil microorganisms.

Such ideas concerning decomposable polymers have appeared in the period
of high economic growth, when consumption of a large amount of polymers was
allowed. They have then been combined with the policy against environmental
pollution. Today, a great amount of PE and PP is produced, and the amount of
abandoned plastics is enormous. Thus, the advent of decomposable polymers has
been long awaited. Here, decomposable polymers are thought of as being polymers
that possess the following properties: during a certain period, the polymers are
stable, then they are rapidly decomposed. Almost all *natural* polymers are
decomposed by natural processes. Composite materials obtained from vinyl
polymers, in which sugar (hemiacetal) or peptide (amide bonding), or the
polymers which have reactive bonds in main chain are also being developed.
However, it is doubtful whether the various requirements for polymer materials
from an engineering and manufacturing point of view and the requirements for
decomposable polymers which disappear after the use can both be realized
satisfactorily. The secondary pollution resulting from the products of the
decomposition would give rise to a serious problem. In spite of intensive efforts

in basic research, discarded polymers have been treated by incineration.

Recently, many attempts have been made to use polymers as reaction promoters in synthetic chemistry. In this field, polymers are used as follows:

(1) as reactive reagents,
(2) as catalysts,
(3) as protective groups.

(1) and (2) are concisely represented by the scheme shown in Fig. 2.9. If we consider the reaction which produces A–B from A and B, an active intermediate P–A is first formed in the presence of the polymer P, then this P–A easily reacts with B to complete the reaction. When P–A is prepared prior to the reaction it is often called a polymeric reagent or, in some cases, a polymeric catalyst.

The compound [108] obtained from the reaction between PS and hydroquinone (HQ) is one of the typical oxidation–reduction resins, and is also a mild reducing reagent. The reaction from low molecular weight compounds proceeds in a homogeneous system, and a reaction mixture is obtained which contains HQ and oxidized quinone (Q), substrate and the reduced products. In this case it is difficult to separate the desired products from the reaction mixture. By immobilization of the oxidation–reduction functional groups onto the polymer, namely by using the insoluble compound [108], Q and HQ polymers can be easily separated from the reaction mixture. We wish to have polymeric reagents

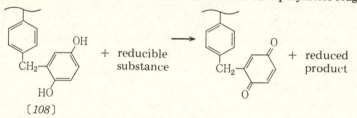

[108]

that can be recovered by appropriate treatments after the reaction and can be used repeatedly. From this point of view, an oxidation–reduction system that is essentially reversible is the most suitable one. Lipoic acid polymer [87] is reversible between thiol (SH) and disulphide (S–S). Polymer-type perbenzoic acid [109] is more easily handled than low molecular weight compounds. However, it is not always easy to recover [109] by a reoxidation of –COOH. The sulphide-type polymer [110] itself is a stable compound; oxidized compounds (aldehyde, ketone) are then obtained by treatment with alcohol and chlorine. It is speculated that the reaction proceeds according to the following reaction scheme in which the sulphide-type polymer is reactivated by chlorine. The

Fig. 2.9.

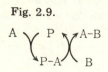

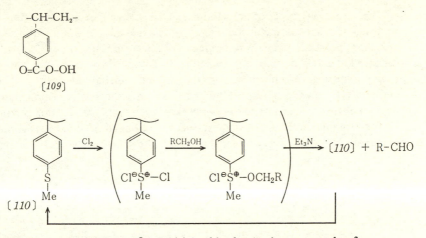

-CH-CH₂-

O=C-O-OH

[109]

[110]

chlorinated nylon described in §2.2.1(e) in this chapter is an example of a polymeric chlorinated reagent, and the formed –NH– type nylon can be used repeatedly after re-chlorination. As previously described, the ester [22] in §2.2.1(b) in this chapter is available as an acyl reagent after the activation process.

Recently, applications for polymers as condensation reagents are also being developed. The carbodiimide structure, which is an acyl activation reagent, reacts with carboxylic acid and amine (amino acid) at room temperature to produce an amide (peptide). Then, the polymer-type Wittig reagent [113] is useful in the synthesis of olefins from corresponding ketones. The problem in this system is the difficulty of reproducing the compounds [111] and [113] from the urea-type polymer [112] and phosphine oxide-type polymer [114], respectively.

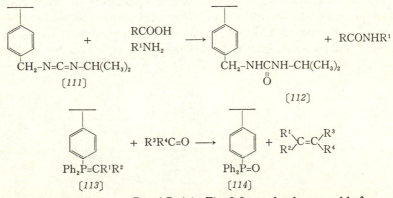

In all cases mentioned above, P and P–A in Fig. 2.9 are clearly separable from each other, and the reactions are stoichiometric. In another case, the reaction is promoted in the presence of a small amount of P or P–A; namely, P or P–A becomes a polymeric catalyst.

There are many types of polymeric catalysts, i.e. acid or base type (ion-exchange resins, etc.), metal complex type, enzyme model and immobilized enzyme type, semiconductor type, photosensitive type and so on. They will be discussed in detail in Chapter 3. Design and analysis of polymeric catalysts have been progressing, together with development in understanding of the reactions and catalysts. Polymeric reagents and catalysts are expected to demonstrate additional virtues, apart from their easy separation from the reaction mixture.

For example, in the following reaction, the obtained ratio for stereoisomers remarkably depends on the molecular weight of the catalyst used. In the reaction

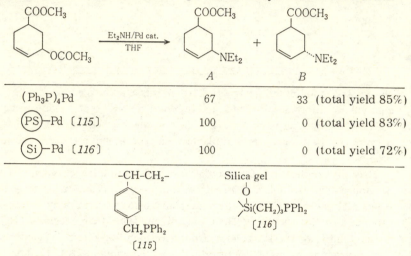

	A	B
$(Ph_3P)_4Pd$	67	33 (total yield 85%)
PS–Pd [115]	100	0 (total yield 83%)
Si–Pd [116]	100	0 (total yield 72%)

–CH–CH₂–

CH₂PPh₂

[115]

Silica gel

Si(CH₂)₃PPh₂

[116]

with the polymeric catalysts, namely triphenyl phosphine type Pd-catalyst combined with Pd–PS (crosslinked), or silica gel ([115], [116]), the product A is predominant, though the mixture of A and B is produced in the reaction with the low molecular Pd-complex. B. M. Trost (1978) has proposed the following reaction mechanism. The reaction proceeds via two reaction paths, as shown by (a) and (b). In the case of the polymeric catalyst, the reaction path (b) which

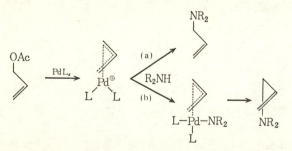

includes an attacking of amine against Pd is sterically inhibited. Therefore, the formation of the substituted product (B) which is formed by an attack from the Pd-side, is suppressed.

Recently, from a standpoint similar to phase transfer catalyst, the following nucleophilic substitution reaction was performed in benzene/water containing anion exchange resin. The term triphase catalyst (organic, water and resin) is proposed.

$$n\text{-}C_8H_{17}Br + KCN \xrightarrow[\text{(PS)}-CH_2N^{\oplus}(CH_3)_3Cl^{\ominus}]{} n\text{-}C_8H_{17}CN + KBr$$

Hetero-cycle formation via the following process has been carried out by using solid phase polymeric reagents; carbanion formation from acetophenone, benzoylation by the activated acyl reagent, and the ring closure of the obtained product by treatment with polymeric hydrazine salt. This process is based on the principle that a species of low molecular weight in liquid phase mediates the reaction between the solid phases which never occurs without the liquid phase.

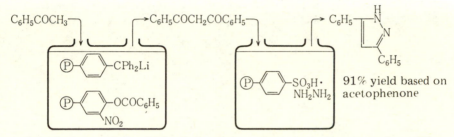

91% yield based on acetophenone

A typical example in which a polymer is used as a protective group, is peptide synthesis in solid phase, which was discovered by R. B. Merrifield (1963). The activation of COOH groups in amino acids by esterification was mentioned already. Here, the amino acid is fixed onto the polymer by esterification, which is followed by the condensation of a terminal amino group in the polymer with a free COOH group of an amino acid in the solution. Thus peptides with desired length and sequence are obtained on the polymer. Finally, the peptide is separated from the polymer. This process is carried out in a flow system, which is different

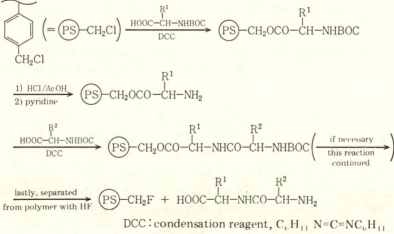

DCC : condensation reagent, $C_6H_{11}N{=}C{=}NC_6H_{11}$
BOC : NH$_2$ protective group, t-BuOCO—

from a solution method using a low molecular weight protective group. The process does not involve separation and purification. Therefore, when many amino acids are treated, experiments can be done in remarkably short periods. Here, PS–CH$_2$Cl is a reactive polymer, and then the protective ester group must possess an appropriate stability as well as a reactivity to cleave that bond. From this point of view, various supporting polymers and reactive groups other than PS–CH$_2$Cl have been developed.

Functional compounds other than amino acids are used in processes based on this idea, i.e. one functional group in the compound is immobilized onto the polymer. Nucleosides [117] and nucleotides are examples, and oligonucleic acids are formed in such a way that the 5′-OH group is bonded onto reactive polymers like [118] and protected, and then the propagation starts from the 3′-OH. Acetylenic alcohol [121], an important raw material for a sex attractant,

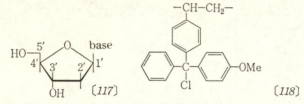

is synthesized by using compound [118] (abbreviated as PS–Cl) as a protective group. It is an important feature of the reaction on the solid polymer that only one reactive group in the diol can react in the step (118) → (119). The com-

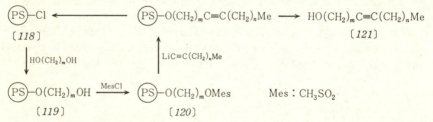

pound [122] is a useful reactive polymer which protects two specific OH groups in sugars, and is easily separable after the reaction, because the hydrolysis of boric ester easily occurs.

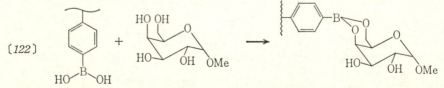

Thus, remarkable progress in the application of reactive polymers in synthetic chemistry is found even in the specialized areas of chemistry, such as preparation of chlorophyll, haem and immune bodies. This technique, with great interest and potential, is expected to develop rapidly.

3

MACROMOLECULES AS CATALYSTS

A   SIMPLE MACROMOLECULAR CATALYSTS
NORIO ISE

3.1   Introductory remarks

In biological systems, the enzymes catalyse biochemical reactions very efficiently and selectively under physiological, mild conditions. In this chapter, detailed discussion of enzymes themselves is not given but emphasis will be placed on catalytic functions of synthetic macromolecules. From both basic and practical points of view, it would be quite interesting if the properties and functions of these fascinating, naturally-occurring biocatalysts could be imitated, even if only partially, by synthetic macromolecules.

We deal in this chapter with the recent study of synthetic catalytic macromolecules and their functions. In the case of enzymes, several different kinds of interaction occur between substrate and enzyme – in a definite order – and create superior catalytic activity. Since these interactions are caused by various chemical groups in the enzymes and substrates, it is quite logical to try to prepare synthetic macromolecules containing these groups for the present purpose. However it is a formidable task, and even more so if we wish to introduce these groups into the synthetic macromolecules with a specific order and spacing. Thus in the study of synthetic catalytic macromolecules carried out up to now, relatively simple reactions have been chosen and one or two of these interactions have been discussed. These studies have revealed that there are cases where synthetic macromolecules can display fairly satisfactory catalytic functions, although not comparable to what enzymes can demonstrate.

In the first part of this chapter we discuss the catalytic systems which involve a single interaction. In such simplified cases, synthetic macromolecules, which have very 'naïve' functions, quite unlike enzymes, are considered. However, by employing these macromolecules, it is possible to dissect the general fundamentals of macromolecular catalysis. In the second part we proceed further to more complicated cases.

3.2   Catalysis by electrostatic interaction

Electrostatic interaction is of a more long-range nature than other interactions. It usually produces the most pronounced effects in catalysis, although this statement is not always true. For example, when synthetic macro-molecules carry many carboxylate or sulphonate groups (i.e. when the macro-molecules are polyelectrolytes), and when substrates are also electrically charged, remarkably high rates of catalysis may be expected. Thus, intensive study has been directed to how interionic reactions are affected by macroions. Several representative cases will be discussed below.

3.2.1   *Reactions between reactants carrying charges of the same sign*
         The following two reactions belong to this category.

$$CH_2BrCOO^\ominus + S_2O_3^{2\ominus} \rightarrow CH_2(S_2O_3)COO^{2\ominus} + Br^\ominus \qquad (3.1)$$

$$2[Co(NH_3)_5Br^{2\oplus}] + Hg^{2\oplus} + 2H_2O \rightarrow 2[Co(NH_3)_5H_2O^{3\oplus}] + HgBr_2 \quad (3.2)$$

It has been pointed out that practically no side reactions are involved. Intensive investigation was performed on these reactions in the 1920s in connection with the so-called salt effect. The influence of polyelectrolytes on reactions (3.1) and (3.2) are shown in Figs. 3.1 and 3.2, respectively. Obviously, the reaction (3.1) is accelerated by the addition of polyethylenimine hydrochloride (PEI · HCl) [1], whereas it is hardly affected by sodium polyacrylate (NaPAA) [2]. On the other

Fig. 3.1. Influence of polyelectrolyte and simple salts on the $CH_2BrCOO^\ominus$–$S_2O_3^{2\ominus}$ reaction. The reactant concentration is 0.01M, unless otherwise specified by the curves, and the temperature is 25 °C. (N. Ise & F. Matsui, *J. Amer. Chem. Soc.*, **90**, 4242 (1968).)

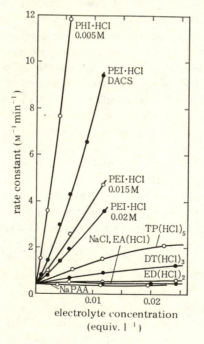

hand, the reaction (3.2) between cations is greatly accelerated by sodium polystyrenesulphonate (NaPSS) [3] and sodium polyethylenesulphonate (NaPES) [4]. NaPAA demonstrates practically no influence because the reaction was studied at low pHs where polyacrylate was in the undissociated form.

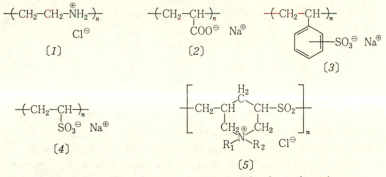

It is well established that salts of low molecular weight show the primary salt effect for these reactions. As a matter of fact, the reaction rate is increased by addition of sodium chloride or sodium sulphate. However, the degree of enhancement by simple salts is strikingly small compared with that demonstrated by macroions, to such an extent that the simple salt effects cannot be seen in Figs. 3.1 and 3.2. In this respect, the behaviour of oligomeric electrolytes

Fig. 3.2. Influence of polyelectrolytes and simple salts on the $Co(NH_3)_5Br^{2\oplus}$–$Hg^{2\oplus}$ reaction. The experimental conditions are as follows: $[Co(NH_3)_5Br^{2\oplus}] = 8 \times 10^{-5}$ M, $[Hg^{2\oplus}] = 2.67 \times 10^{-4}$ M, $[HNO_3] = 8.02 \times 10^{-4}$ M, temperature = 15 °C. ○: NaPSS, ●: NaPES, x: Na_2SO_4, ▼: PEI · HCl, △: NaPAA, □: NaCl. (N. Ise & F. Matsui, *J. Amer. Chem. Soc.*, **90**, 4242 (1968).)

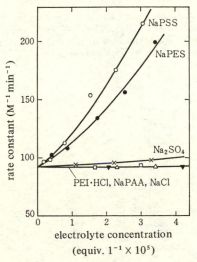

is of interest. When the reaction (3.1) is allowed to proceed in the presence of low molecular weight analogues of PEI · HCl, namely ethylenediamine $(ED(HCl)_2)$, diethylenetriamine $(DT(HCl)_3)$, and tetraethylenepentamine $(TP(HCl)_5)$, the reaction rate becomes larger with increasing degree of polymerization (or molecular weight) of the oligomers. This tendency is clearly seen from Fig. 3.3. It is evident that the higher the molecular weight is, the larger the rate.

Summarizing the above results, we may state:

(i) The reaction between similarly charged ionic species is greatly accelerated by macroions having a charge of the opposite sign.

(ii) The acceleration by high molecular weight compounds is much larger than that by corresponding low molecular weight analogues.

These conclusions were confirmed to be true for other reaction-polyelectrolyte systems. As stated above, the observed acceleration is due to the electrostatic interaction between ionic reactants and the oppositely charged macroions; other interactions do not play a role as far as the reactions under consideration are concerned. This can be understood on the basis of the fact that two cationic macroions, PEI · HCl and a copolymer of diallyldimethylammonium chloride and sulphur dioxide (a compound designated as DACS in Fig. 3.1, with $R_1 = R_2 = CH_3$ in the structural formula [5]), accelerate the reaction approximately to the same extent, when the equivalent concentration is the same. In spite of the different chemical structures and linear charge densities, these two macrocations display quantitatively about the same effect. This would not be the case, if interactions other than electrostatic interaction were key factors.

The strength of the electrostatic interaction depends on the number of charges which the macroions and reactant ions carry. When two reactants of different valencies are compared in the presence of a common macroion,

Fig. 3.3. Molecular weight dependence of the acceleration by polyelectrolyte. The rate constant of the $CH_2BrCOO^{\ominus}$-$S_2O_3^{2\ominus}$ reaction is plotted against the degree of polymerization of polyethylenimine hydrochloride and its low molecular weight analogues.

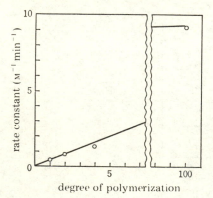

degree of polymerization

a larger effect of the macroion can be expected for reactants of higher valency. For example, the electron-transfer reactions between $Fe^{2\oplus}$ and bivalent complexes such as $Co(NH_3)_5Cl^{2\oplus}$, $Co(NH_3)_5Br^{2\oplus}$ and $Co(NH_3)_5N_3^{2\oplus}$ can be accelerated 10^3-fold by NaPES. On the other hand, at the same concentration, the reactions between $Fe^{2\oplus}$ and monovalent complexes, namely *trans*-$Co(en)_2Cl_2^{\oplus}$ and *cis*-$Co(NH_3)_4(N_3)_2^{\oplus}$ (en stands for ethylenediamine) are speeded up by only a factor of 20. The reaction (3.3) between polyvalent reactants shows 10^4-fold acceleration by NaPES. A cationic polyelectrolyte, polybrene [6], accelerates reaction (3.4) by a factor of 10^5 (see Fig. 3.4).

$$U^{4\oplus} + Tl^{3\oplus} + 2H_2O \rightarrow UO_2^{2\oplus} + Tl^{\oplus} + 4H^{\oplus} \tag{3.3}$$

$$2Fe(CN)_6^{4\ominus} + S_2O_8^{2\ominus} \rightarrow 2Fe(CN)_6^{3\ominus} + 2SO_4^{2\ominus} \tag{3.4}$$

[6]
$$\left[-\underset{\underset{CH_3}{|}}{\overset{\overset{CH_3}{|}}{N^{\oplus}}}(CH_2)_6 - \underset{\underset{CH_3}{|}}{\overset{\overset{CH_3}{|}}{N^{\oplus}}}(CH_2)_3 - \right]_n$$

Fig. 3.4. Acceleration of the $Fe(CN)_6^{4\ominus}$-$S_2O_8^{2\ominus}$ reaction by polybrene at 25 °C. The dashed lines represents the calculated values by the Brönsted–Bjerrum–Manning theory. k_2* is the rate constant at zero concentration of macroion. (A. Enokida, T. Okubo & N. Ise, *Macromolecules*, **13**, 48 (1980).)

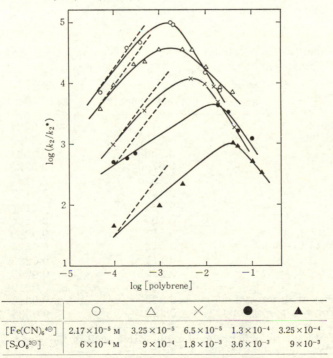

	○	△	×	●	▲
$[Fe(CN)_6^{4\ominus}]$	2.17×10^{-5} M	3.25×10^{-5}	6.5×10^{-5}	1.3×10^{-4}	3.25×10^{-4}
$[S_2O_8^{2\ominus}]$	6×10^{-4} M	9×10^{-4}	1.8×10^{-3}	3.6×10^{-3}	9×10^{-3}

In the above, only simple inorganic reactions have been mentioned. Qualitatively the same tendency was observed for a variety of organic, ionic reactions, although we will not discuss those here to save space. Readers are referred to the reading list given at the end of this book.

3.2.2   Reactions between oppositely charged reactants

The simplest reaction in this category is the formation of urea.

$$NH_4^{\oplus} + OCN^{\ominus} \rightleftharpoons (NH_2)_2CO \tag{3.5}$$

This reaction was earlier found to be decelerated by addition of simple salts. Similarly, polymeric salts are known to retard the reaction. Table 3.1 illustrates some of the results recently obtained. For example, addition of (anionic) NaPAA decelerates the reaction and this is the case with cationic DECS. More interesting is the comparison between polymeric and simple salts. At the same molar concentration, NaPAA gives smaller rate constants than either NaCl or $CaCl_2$. Water-soluble macromolecules such as polyethyleneglycol and polyvinylpyrrolidone have no substantial influence on the rate constant, indicating that the rate decrease demonstrated by NaPAA and DECS is not due to viscosity increment of the solution. From these observations, it may be concluded that:

(iii) Both cationic and anionic macroions have a retarding action on the reactions between oppositely charged reactants.

Table 3.1. *Deceleration of the $NH_4^{\oplus}$-$OCN^{\ominus}$ reaction by polymeric and simple salts at 50 °C.* $[NH_4^{\oplus}] = [OCN^{\ominus}] = 0.1025$ M

Salt	salt concn (M)	k_2 (M^{-1} min^{-1})
none	—	0.0324
NaPAA	0.0200	0.0302
	0.0588	0.0235
	0.0700	0.0225
	0.112	0.0181
	0.150	0.0137
	0.223	0.0123
DECS[a]	0.100	0.0185
$CaCl_2$	0.111	0.0268
NaCl	0.0444	0.0324
	0.111	0.0285
	0.200	0.0246
	0.300	0.0219
Polyethyleneglycol	0.222	0.0313
Polyvinylpyrrolidone	0.222	0.0325

[a] DECS is a cationic macromolecule with $R_1 = R_2 = C_2H_5$ of the structure [5].

(iv) High molecular weight compounds show larger deceleration than low molecular weight ones.

3.2.3   Influence of macroions on interionic equilibrium reactions

The rate, v, or the rate constant, k, which was discussed above, was an apparent value. In all chemical reactions, forward and backward reaction steps are involved. If we denote the respective rates by $\vec{v}$ and $\overleftarrow{v}$, respectively, $v = \vec{v} - \overleftarrow{v}$. Therefore, the assertion that macroions enhance a reaction is not always unambiguous. It has to be made clear which of $\vec{v}(\vec{k})$ or $\overleftarrow{v}(\overleftarrow{k})$ was affected by the macroions. In this respect, it is interesting to study equilibrium reactions. The complex formation of metal ions with murexide (3.6) was studied by the temperature-jump or stopped-flow method.

$$M^{n+} + Mx^{\ominus} \rightleftharpoons (M \ldots Mx)^{(n-1)\oplus} \qquad (M^{n+}: \text{metal ions, Mx: murexide})$$

$$(3.6)$$

The forward reaction proceeds between oppositely charged reactants so that addition of macroions (for example NaPES) retards the reaction, in other words, $\vec{k}/\vec{k}* < 1$, where the asterisk denotes zero concentration of macroions. Experimental results on the Cu^{2+}-Mx reaction are shown in Fig. 3.5. In this figure, the influence of the macroions on the equilibrium constant (K_a) is also given. It is noteworthy that $\vec{k}/\vec{k}*$ falls on the same curve as K_a/K_a^*. Because $K_a = \vec{k}/\overleftarrow{k}$ by definition, the observation mentioned above indicates that the backward step was hardly influenced by the addition of polymer. This fact would be related to the bulkiness of the metal complex, in which the electric charge is delocalized. That the forward and backward steps are affected in

Fig. 3.5. Polyelectrolyte influence on the forward rate constant $\vec{k}$ and the equilibrium constant K_a of the $Cu^{2\oplus}$-Murexide complex formation. The filled symbols are for $\vec{k}/\vec{k}*$ and the open ones for K_a/K_a^*. $[Cu^{2\oplus}] = 2 \times 10^{-4}$ M, $[\text{Murexide}] = 5 \times 10^{-5}$ M, pH 4, $[\text{NaCl}] = 0.05$ M; 25 °C. For comparison, the micellar effects by sodium lauryl sulphate (NaLS), sodium decylsulphate (NaDS) and sodium octylsulphate (NaOS) are shown in addition to the NaPES effect. (N. Ise, *J. Polymer Sci., Polymer Symposia*, **62**, 205 (1978).)

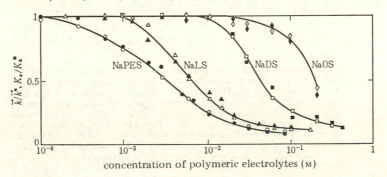

concentration of polymeric electrolytes (M)

a different manner by macroions was further confirmed to be true for other equilibrium reactions. Thus we can state that:

(v) Macroions shift the chemical equilibria; in other words, they affect the forward and backward steps independently.

This observation leads us to the assertion that macroions should not be called catalysts. The reason is simply because catalysts have been defined by Wo. Ostwald as substances that affect reaction reates without shifting chemical equilibria. In other words, catalysts alter the rates of the forward and backward steps in the same proportion. If this definition is strictly applied, macroions are to be regarded simply as reaction promoters or retarders. On the other hand, it has to be remembered that most enzymes are catalysts with high turn-over numbers, and accelerate simultaneously forward and backward steps. Though fairly strong acceleration effects by synthetic macromolecules have been found, their mechanism of acceleration has to be understood as basically different from that of the enzymes.

3.2.4 *Thermodynamics of acceleration and deceleration by macroions*

The next problem to dissect is how the acceleration on the interionic reactions between similarly charged reactants and the deceleration on those between oppositely charged reactants are brought about. In this connection, we discuss here the free energy of activation ($\Delta G^{\neq}$). According to the transition-state theory, the second-order rate constant k_2 of the reaction (3.7) can be written as equation (3.8),

$$A^{z_A} + B^{z_B} \rightleftharpoons [AB]^{z_A + z_B} \rightarrow C + D \tag{3.7}$$

$$k_2 = (kT/h) \exp(-\Delta G^{\neq}/RT)$$
$$= (kT/h) \exp(\Delta S^{\neq}/R) \exp(-\Delta H^{\neq}/RT) \tag{3.8}$$

where k, T, and h are the Boltzmann constant, temperature and Planck constant, and $\Delta S^{\neq}$ and $\Delta H^{\neq}$ are the entropy and enthalpy of activation, respectively. From the temperature dependence of k_2 and equation (3.8), we can estimate $\Delta G^{\neq}$, which is the difference in the free energies of the transition state and reactants. From $\Delta G^{\neq}$, $\Delta H^{\neq}$ and $\Delta S^{\neq}$ can be further determined.[1]

Table 3.2 gives the observed values of the activation quantities for reaction (3.2), which was accelerated by macroions. In Tables 3.3 and 3.4 the observed activation quantities are given for reaction (3.5), which was decelerated, and

---

1 To be exact, the activation quantities on the right-hand side of equation (3.8) should be regarded as 'apparent' in the present case. In systems that contain only simple ions, an equation formally identical to equation (3.8) holds for the standard activation quantities. However, in systems containing macroions, the standard values cannot be determined because interaction between macroions and reactant ions does not vanish even at the infinite dilution of the reactants. In other words, the activity coefficient of the reactant cannot be claimed to be unity at the infinite dilution of the reactants, if macroions coexist. Thus, we have to be satisfied with the apparent values of the activation quantities.

for one of the equilibrium reactions as follows:

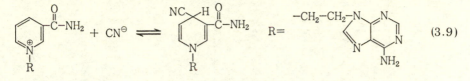

$$(3.9)$$

Needless to say, the acceleration is due to a decrease in $\Delta G^{\neq}$, whereas the deceleration stems from an increase in $\Delta G^{\neq}$. The parameters given for the equilibrium reaction (Table 3.4) require some comment. It has been confirmed experimentally that the forward step of this reaction is retarded by addition of polyelectrolyte, which is reflected in an increase in $\Delta G^{\neq}$. Furthermore the equilibrium constant $K (= \overrightarrow{k}/\overleftarrow{k})$ has been found to decrease when the polyelectrolyte is added; in other words, $\overleftarrow{k}$ does not vary with the polyelectrolyte addition. The increase in ΔG obviously corresponds to the shift of K; and $\Delta G_{-}^{\neq}$, which is equal to $(\Delta G^{\neq} - \Delta G)$ and represents the free energy of activation

Table 3.2. *Activation parameters of the polyelectrolyte-accelerated* $Co(NH_3)_5Br^{2\oplus}$–$Hg^{2\oplus}$ *reaction*

Electrolyte added	conc. $(10^{-5} \times$ equiv. $l^{-1})$	$\Delta G^{\neq}$ (kJ mol^{-1})	$\Delta H^{\neq}$ (kJ mol^{-1})	$\Delta S^{\neq}$ (J K^{-1} mol^{-1})
—	—	70	50	−68
Polystyrene-	0.772	69	41	−97
sulphonate	2.31	68	26	−147
Polyethylene-	0.860	69	45	−82
sulphonate	2.58	68	31	−129
Sodium chloride	3.00	70	50	−68
Sodium sulphate	3.31	69	50	−67

15 °C; $[Co(NH_3)_5 Br^{2\oplus}] = 8 \times 10^{-5}$ M, $[Hg^{2\oplus}] = 2.67 \times 10^{-4}$ M, $[HNO_3] = 8.02 \times 10^{-4}$ M.

Table 3.3 *Activation parameters of the polyelectrolyte-decelerated* $NH_4^{\oplus}$–$OCN^{\ominus}$ *reaction*

Electrolyte added	$\Delta G^{\neq}$ (kJ mol^{-1})	$\Delta H^{\neq}$ (kJ mol^{-1})	$\Delta S^{\neq}$ (J K^{-1} mol^{-1})
—	98.7	95	−8.4
Sodium polyacrylate	102	83	−46
Calcium chloride	100	83	−54
Sodium chloride	99	87	−33

Electrolyte conc. = 0.111 M, $[NH_4^{\oplus}] = [OCN^{\ominus}] = 0.0513$ M; 50 °C.

Table 3.4. *Thermodynamic and activation parameters of the nicotinamide-derivative–CN^- reaction*

Electrolyte added	$\Delta G^{\neq}$ (kJ mol^{-1})	$\Delta H^{\neq}$ (kJ mol^{-1})	$\Delta S^{\neq}$ (J K^{-1} mol^{-1})	ΔG (kJ mol^{-1})	ΔH (kJ mol^{-1})	ΔS (J K^{-1} mol^{-1})	$\Delta G^{\neq}$ (kJ mol^{-1})
–	73	38	−117	−12.9	−41	−96	85.9
DECS	75.3	34	−138	−10.8	−31	−68	86.1
Sodium polyethylene sulphonate	75.7	39	−121	−10.5	−32	−71	86.2
Sodium polystyrene sulphonate	75.3	38	−121	−10.9	−28	−59	86.2

DECS: polymeric cations with $R_1 = R_2 = C_2H_5$ in the structural formula [5].
$\Delta G_-^{\neq} = \Delta G^{\neq} - \Delta G$, [nicotinamide derivative] = 1.98×10^{-3} M, [KCN] = 4.0×10^{-3} M, [KOH] = 1.0×10^{-3} M, [DECS] = [NaPES] = 2×10^{-2} equiv l^{-1}, [NaPSS] = 2×10^{-3} equiv l^{-1}.

of the backward step, stays constant ($86\,\mathrm{kJ\,mol^{-1}}$) with or without the poly-electrolytes and irrespective of the nature of polyelectrolytes, corresponding to the observed fact that k was not affected by polyelectrolytes.

$\Delta G^{\neq}$ represents by definition the difference between the free energies of the activated complex and that of the reactants. Therefore, the decrease in $\Delta G^{\neq}$ implies two possibilities, as is shown in (1-a) and (1-b) of Fig. 3.6. Similarly, the $\Delta G^{\neq}$ decrease is due to either of two possibilities shown in (2-a) and (2-b). Which of (a) and (b) is responsible in reality is an interesting question and can be answered, though qualitatively, as follows. Brønsted and Bjerrum derived equation (3.10) for reaction (3.7),

$$k_2/k_{20} = f_A \cdot f_B/f_{AB} \tag{3.10}$$

where k_{20} is the rate constant at the zero ionic strength and f is the activity coefficient. It is a well-established matter in basic physical chemistry that the primary salt effect, namely the rate constant change of interionic reactions with varying ionic strength, can be accounted for most successfully by equation (3.10) combined with the Debye–Hückel theory. According to the existing data, the activity coefficient of an ionic species becomes smaller with increasing ionic strength as far as very dilute solutions are concerned, and the larger the valency of the ionic species, the more significant the decrease of the activity coefficient. Since the valency of the activated complex can be considered to be equal to the *algebraic* sum of the valencies of the reactants, the activity coefficient of the activated complex f_{AB} might be smaller or larger than the

Fig. 3.6. Free energy diagram of polyelectrolyte catalysis. (1) Accelera-tion; (2) deceleration; (3) the forward process is retarded whereas the backward process is unaffected.

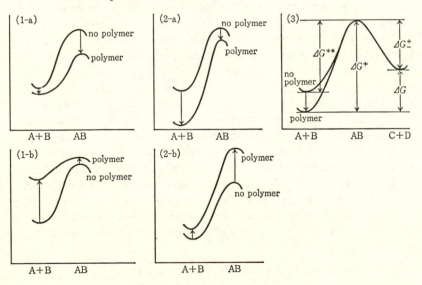

product $f_A \cdot f_B$ for reactions between similarly charged ion species, and for those between oppositely charged species. Therefore the left-hand side of equation (3.10), namely k_2/k_{20}, would become larger than unity for the former reactions. In other words, acceleration would result when the ionic strength is raised. The situation is opposite for the latter reactions because k_2/k_{20} is smaller than unity; deceleration can be observed.

The Debye–Hückel theory, which was needed for evaluation of f_A, f_B and f_{AB}, is valid only when simple ionic species are involved. If polyelectrolytes are introduced, this theory has to be abandoned, although equation (3.10) continues to be valid because of its general character. Instead, the use of a theory for the evaluation of the activity coefficient terms in polyelectrolyte solutions, proposed by Manning, was found to be fairly satisfactory for description of the variation of k_2/k_{20} with polyelectrolyte addition. Although detailed discussion on the theoretical analyses will not be given here, it may be claimed that the rate enhancement and retardation by polyelectrolytes, though overwhelmingly larger than those demonstrated by simple electrolytes, are basically primary salt effects. The apparent difference between the two types of electrolyte is due to the lowering of the activity coefficients of reactant ions being more pronounced in the case of polyelectrolytes than simple electrolytes, as has been experimentally confirmed.[1]

The above consideration shows that the situation presented in (1-a) of Fig. 3.6 is correct in reality for the accelerated cases. In other words, the activated complex is more strongly stabilized than the reactants by macroions. Similarly, (2-a) can be justified for the decelerated cases; the reactants are stabilized by macroions.

For the equilibrium reaction also, the situation is straight-forward. On the basis of the experimental fact the $\Delta G_-^{\neq}$ did not vary (Table 3.4), it may be assumed that the free energy levels of the activated complex and the product are not dependent on whether the polyelectrolyte is added or not, as is shown in (3) of Fig. 3.6. Then what was changed by addition of the polyelectrolyte is the free energy of the reactants. It decreased so that $\Delta G^{\neq}$ was increased; in other words the forward rate constant was lowered.

3.2.5 *Catalysis by micelles and electrically charged polymer latex particles*

The acceleration or deceleration exhibited by polyelectrolytes was brought about by the electrostatic interaction between the macroions and activated complex or reactants. It would easily be accepted that the strong interactions are due to the numerous electric charges of the macroions being confined by covalent bonds in a rather small space. This confinement is possible not only by covalent bonds but also by other kinds of attraction. For example, detergent micelles have a concentrated charge distribution in the neighbourhood of the micelle domain as a consequence of hydrophobic interaction between

1 See, for example, N. Ise, *Adv. Polymer Sci.*, **7**, 536 (1971).

hydrocarbon chains. A similar situation is encountered with electrically charged polymer latex particles. By using the emulsion polymerization technique, polymerizations of styrene with potassium persulphate as initiator give spherical particles comprising polystyrene as 'core' and ionized groups as 'skin'. Thus both detergent micelles and polymer latex particles exhibit catalytic influence similar to macroions. The micellar effects on the equilibrium reaction (3.6) were demonstrated in Fig. 3.5. It is noteworthy that the effect comes into existence only when the surfactant concentration is above the so-called critical micelle concentration (10^{-3} M for sodium lauryl sulphate). It must be recalled that the 'concentration' of charges takes place when micellization occurs. The negatively charged polystyrene particles have an accelerating effect on reaction (3.2), as is shown in Fig. 3.7. The basic points are clearly seen, although the relative merits of the three charged solutes cannot be decided because the charge number of each solute was not the same. Fig. 3.8 gives another example of the influence of styrene–acrylate latex particles on a reaction.

$$Co(NH_3)_5Br^{2\oplus} + OH^{\ominus} \rightarrow Co(NH_3)_5OH^{2\oplus} \tag{3.11}$$

The reaction between oppositely charged reactants is retarded as in the case of soluble polyelectrolytes such as NaPAA. The retardation seems to be dependent on the surface charge density, but neither on the acrylate content in the particles nor on their diameter.

3.2.6 *Mechanism of stabilization by macroions: volume of activation and solvation–desolvation*

As was mentioned above, the acceleration is due to the stabilization of the activated complex whereas the deceleration is brought about by the stabili-

Fig. 3.7. Acceleration by polyelectrolyte (NaPES), polystyrene latex particle (PS-Latex) and detergent (NaLS) on the $Co(NH_3)_5Br^{2\oplus}$-$Hg^{2\oplus}$ reaction. $[Co(NH_3)_5Br^{2\oplus}] = 1.8 \times 10^{-3}$ M, $[Hg^{2\oplus}] = 4.2 \times 10^{-4}$ M $[H^+] = 10^{-3}$ M; 25 °C. (N. Ise & M. Ishikawa, presented at the 24th Annual Meeting of the Society of High Polymers, Japan, 1975, Tokyo, Preprint p. 77.)

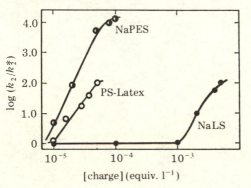

zation of the reactants. The factor causing the stabilization is, at relatively lower polymer concentrations, the electrostatic interaction of macroions with the activated complex or reactants.[1] However, at high concentrations of polymers, or for fairly large acceleration, the stabilization appears to be caused by another factor, in addition to the electrostatic factor. In this respect, the influence of high pressure on polyelectrolyte-accelerated reactions has been studied on reaction (3.11) and reactions (3.12) and (3.13).

$$Co(NH_3)_5Br^{2\oplus} + H_2O \rightarrow Co(NH_3)_5H_2O^{3\oplus} \qquad (3.12)$$

$$Co(NH_3)_5Br^{2\oplus} + Ag^{\oplus} + H_2O \rightarrow Co(NH_3)_5H_2O^{3\oplus} + AgBr \qquad (3.13)$$

The reactions (3.11), (3.12) and (3.13) are between oppositely charged reactants, spontaneous aquation and Ag^+-induced aquation (between similarly charged reactants), respectively.

The volume of activation ($\Delta V^{\neq}$) is given by equation (3.14) and is equal to

$$(\partial \ln k/\partial P)_T = -\Delta V^{\neq}/RT, \qquad (3.14)$$

the difference between the partial molar volume of activated complex and that of reactants. Furthermore, $\Delta V^{\neq}$ is divided into two parts by equation (3.15),

$$\Delta V^{\neq} = \Delta V_1^{\neq} + \Delta V_2^{\neq}. \qquad (3.15)$$

where $\Delta V_1^{\neq}$ represents the volume change due to changes of interatomic

1 For this reason, the catalysis by polyelectrolyte could be accounted for in terms of the Brønsted–Bjerrum equation with Manning's theory on polyelectrolyte solutions, which takes into consideration the electrostatic interaction only, when the polyelectrolyte concentration is low.

Fig. 3.8. Alkaline hydrolysis of $Co(NH_3)_5Br^{2\oplus}$ in the presence of negatively charged polymer latex particles at 25 °C. $[Co(NH_3)_5Br^{2\oplus}]=$ 1.43×10^{-5}M, $[OH^{\ominus}] = 7.79 \times 10^{-3}$M. The abscissa is given in mol l^{-1} of the carboxyl groups. The latex particles were prepared by emulsion copolymerization of styrene and acrylate. The diameter of the particles ranges from 5100 Å (S-8) to 6800 Å (S-2) and the surface charge density is in a range between 0.0087 Å^{-2} (S-6) and 0.14 Å^{-2} (S-4).

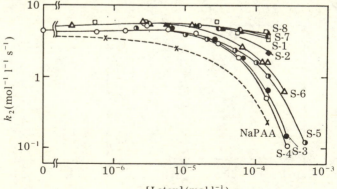

Table 3.5. *Activation parameters of the $Co(NH_3)_5Br^{2\oplus}-Y$ reaction and influence of polyelectrolytes thereon*

Reaction	Y	$k/k*$	$\Delta V^{\neq}*$	$\Delta V^{\neq}$	$\Delta S^{\neq}**$	$\Delta S^{\neq}$	$\Delta H^{\neq}**$	$\Delta H^{\neq}$	$\Delta G^{\neq}**$	$\Delta G^{\neq}$
(3.11)	OH^-	0.045^a	+32	+10	+117	+75	104	98	68	76
(3.12)	H_2O	2^a	−8.7	−0.2	−63	−16	84	96	103	101
(3.13)	$Ag^{\oplus} + H_2O$	90^b	−5.3	+30	−180	+22	33	82	86	75

$\Delta V^{\neq}$, in ml mol^{-1}, $\Delta S^{\neq}$ in J K^{-1} mol^{-1}, $\Delta H^{\neq}$ and $\Delta G^{\neq}$ in kJ mol^{-1}.
^{a} [NaPES] = 10^{-4} equiv l^{-1}, ^{b} [NaPSS] = 5 × 10^{-5} equiv l^{-1}.
The asterisks indicate zero polymer concentration.

distance in the course of activation and $\Delta V_2^{\neq}$ is the difference between the solvation states of the activated complex and that of the reactants. As far as interionic reactions are concerned, the second contribution ($\Delta V_2^{\neq}$) is known to be much more significant than $\Delta V_1^{\neq}$. In other words, $\Delta V^{\neq}$ can be attributed exclusively to the volume factor caused by changes in the solvation state.

As was found earlier, the volume of solution contracts by $20\ ml\ mol^{-1}$, when water dissociates, or for $2H_2O \rightarrow H_3O^{\oplus} + OH^{\ominus}$. This contraction is caused by strong hydration around newly produced ionic species. Generally speaking, the volume contraction by solvation of ions from 1-1 type electrolytes is in the range 10–$30\ ml\ mol^{-1}$ in water and organic solvents. With this value in mind, the $\Delta V^{\neq}$ values of the reactions (3.11), (3.12) and (3.13) shown in Table 3.5 at zero polymer concentration (i.e., $\Delta V^{\neq *}$) may be interpreted as follows. Reaction (3.11) is the one between oppositely charged reactants. Thus, the activated complex may be assumed to be monovalent, whereas the reactant $Co(NH_3)_5Br^{2\oplus}$ is bivalent. Thus, relative lowering of the valency causes dehydration from the activated complex so that the solution volume increases, resulting in a positive $\Delta V^{\neq *}$. On the other hand, for reaction (3.12), the activated complex is believed to be $Co(NH_3)_5^{3\oplus}$ which is produced by dissociation of $Br^{\ominus}$. The increase in the valency in the course of activation causes stronger hydration for the activated complex than for the reactant, so that $\Delta V^{\neq *}$ is negative. In order to facilitate understanding, a highly simplified scheme is presented in Fig. 3.9

Fig. 3.9. Simplified scheme for hydration of ionic reactants and activated complex and dehydration by macroions. (a) without macroions, (b) with macroions. It should be noted that the figure is drawn for reactions between similarly charged reactants. (N. Ise, T. Maruno and T. Okubo, *Proc. Royal. Soc., Lond.*, **A370**, 485 (1980).)

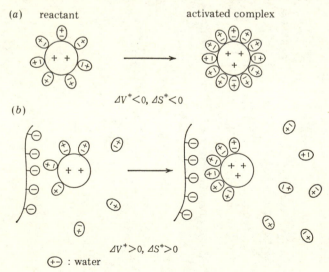

for hydration and dehydration of ionic solutes. The ion–dipole interaction between the ionic solute and water is intensified as the reaction proceeds from the reactant to the activated complex (in the case of the reaction between similarly charged ionic species). The activated complex of reaction (3.13) is also a trivalent cation, being more strongly hydrated than the reactant. Thus, we have a negative $\Delta V^{\neq *}$.

The addition of negatively charged macroions into these reaction systems causes deceleration for reaction (3.11) and acceleration for reactions (3.12) and (3.13). It is worth mentioning that $\Delta V^{\neq}$ changed with the addition of polyelectrolytes. For reaction (3.11), $\Delta V^{\neq *}$ was $+32\,\mathrm{ml\,mol^{-1}}$, reflecting dehydration in the course of activation. The deceleration by the anionic macroion is accompanied by the decrease in $\Delta V^{\neq}$ from $+32$ to $+10\,\mathrm{ml\,mol^{-1}}$. Since $OH^{\ominus}$ can be repelled by macroions and the monovalent activated complex is less strongly attracted by the macroions than the bivalent Co-complex, this decrease indicates that the reactant, namely $Co(NH_3)_5Br^{2\oplus}$, is dehydrated by the macroion, which is due to weakening of the ion–dipole (ion–water) inter-action, subject to the intense electrostatic field of oppositely charged macroions. Correspondingly, $\Delta S^{\neq}$ decreases with addition of the macroions from $+117$ to $+75\,\mathrm{J\,K^{-1}\,mol^{-1}}$. Also, $\Delta H^{\neq}$ decreases from $+104$ to $+98\,\mathrm{kJ\,mol^{-1}}$. Thus the deceleration is really entropic and is due to stabilization of the reactant, in line with (2-a) of Fig. 3.6.

For reaction (3.12), $\Delta V^{\neq}$ increased from -8.7 to $-0.2\,\mathrm{ml\,mol^{-1}}$ by addition of NaPES. The increase implies that the activated complex is dehydrated by the anionic macroions, once again if $\Delta V^{\neq}$ is attributable to the solvation factor only. Similarly, the $\Delta V^{\neq}$ value of reaction (3.13) increased from -5.3 to $+30\,\mathrm{ml\,mol^{-1}}$ with addition of NaPSS, which can be accounted for in terms of the dehydration of the activated complex. For both reactions (3.12) and (3.13), note that $\Delta S^{\neq}$ increases correspondingly by addition of macroions. The enthalpy of activation $(\Delta H^{\neq})$ increases for these two reactions, which is in line with the dehydration of the activated complex. Obviously, the rate increase (or decrease in $\Delta G^{\neq}$) is due to the increase in $\Delta S^{\neq}$, which resulted from the dehydration of the activated complex. These thermodynamic considerations substantiate the free energy diagram shown in (1-a) of Fig. 3.6, but not (1-b).

Summarizing the above discussion, we state that:

(vi) Dehydration of the activated complex by macroions plays an important role in observed acceleration of the reaction between similarly charged reactants. Also, dehydration of the reactant is a key factor in the poly-electrolyte deceleration of the reaction between oppositely charged reactants.

A few words appear to be necessary on the frequently alleged interpretation of macromolecular catalysis. It has been claimed that reactants are accumulated in the vicinity of the macromolecules by electrostatic and/or hydrophobic inter-action so that the collision frequency is increased, resulting in acceleration. The local concentration of reactants can be raised undoubtedly but we question that it does necessarily mean enhanced collision. As a matter of fact, $\Delta S^{\neq}$ increased by addition of macroions for reaction (3.12) and (3.13) and the increase may be taken as implying enhanced collision. However, it cannot be overlooked that the $\Delta S^{\neq}$ increase is associated with a $\Delta V^{\neq}$ increase, or dehydration of the activated complex. In other words, if the interpretation customarily accepted is correct, the relatively large change in $\Delta V^{\neq}$ cannot be understood.

3.3   Catalysis by electrostatic and hydrophobic interactions

Hydrophobic interaction is based on 'attraction' between hydrophobic molecules which comes into existence as a result of strong water–water interaction. When hydrophobic molecules are introduced into water, the iceberg structure due to hydrogen-bond formation is stabilized so that the entropy and enthalpy of the system are decreased. The decrease in entropy is not favourable in the light of the second law of thermodynamics so that a force to raise the entropy would be created. Schematically speaking, a plural number of hydrophobic molecules get close together to stabilize the iceberg structure jointly. The larger the hydrophobic solute, the smaller the entropy of the system and the stronger the hydrophobic interaction. Under some conditions, the hydrophobic interaction exceeds the electrostatic interaction. For example, surfactant ions form micelles above the critical micelle concentration, in spite of the electrostatic repulsion between ionic head groups. If hydrophobic groups are introduced into macromolecules and reactants, chemical reaction has to be influenced quite remarkably. This was earlier confirmed experimentally for macromolecular catalysis of the following reaction. This reaction is catalysed

$$\left\langle\!\!\!\bigcirc\!\!\!\right\rangle\text{-}COOC_2H_5 + H_2O \xrightarrow{+H^{\oplus}} \left\langle\!\!\!\bigcirc\!\!\!\right\rangle\text{-}COOH + C_2H_5OH \qquad (3.16)$$
$$\overset{|}{NH_3^{\oplus}} \qquad\qquad\qquad \overset{|}{NH_3^{\oplus}}$$

by protons and accelerated by addition of HPSS. This acceleration is obviously due to electrostatic interaction between PSS anions and substrate cations and, in addition, hydrophobic interaction between PSS and benzene rings of the substrate.

Various ionic reactions were studied and a few representative cases will be mentioned here. The $S_N Ar$ reaction, such as reaction (3.17), is found to be accelerated by cationic macroions, because it is a reaction between anions. When hydrophobic groups are introduced into the macroions, the acceleration

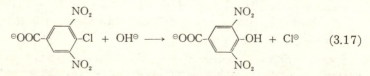

$$(3.17)$$

becomes greater. As is seen from Table 3.6, addition of simple salts slightly enhanced the reaction but the acceleration by the macroions was greater, and

even more so when the macroions became more hydrophobic. The order is
DACS $<$ MBz $<$ MNBz.

In the reaction mentioned above, electrostatic and hydrophobic interactions
are in action, and the results are acceptable when the two interactions have
similar strengths. However, if the hydrophobic interaction is so strong that it
outweighs the electrostatic interaction, an interesting behaviour can be
observed. One example is the alkaline-fading reaction of (hydrophobic) dye

$$R_1-\!\!\!\langle\bigcirc\rangle\!\!\!-C=\!\!\!\langle\bigcirc\rangle\!\!\!=\overset{+}{N}R_3 \ + \ OH^{\ominus} \longrightarrow R_1-\!\!\!\langle\bigcirc\rangle\!\!\!-\underset{OH}{\overset{|}{C}}-\!\!\!\langle\bigcirc\rangle\!\!\!-NR_3 \qquad (3.18)$$

ions (reaction (3.18)). Since this is a reaction of oppositely charged ions, it is
simply to be retarded by both cationic and anionic macroions. However, the
observed facts are more complicated than expected, as is seen from Fig. 3.10
and Table 3.7. First, it is accelerated by hydrophobic, cationic macroions such
as C_{16}BzPVP (copolymer of 4-vinyl-N-benzoylpyridinium chloride and 4-vinyl-N-
n-cetylpyridinium bromide), BzPVP(poly(4-vinyl-N-benzylpyridinium chloride)),
C_4PVP(poly(4-vinyl-N-butylpyridinium bromide)), and C_2PVP(poly(4-vinyl
ethylpyridinium bromide)), which are poly(vinyl pyridine) derivatives of the
structure [7]. Hydrophilic, cationic macroions such as DECS ($R_1 = R_2 = C_2H_5$
in the structure [5]) and hydrophilic anionic macroions such as NaPAA show

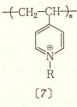

[7]

Table 3.6. *Acceleration of the dinitrochlorobenzoic acid–OH^-
reaction by hydrophobic cationic macroions*

Macroion	Concentration (equiv. l^{-1})	$k_2 \times 10^3$ ($M^{-1} s^{-1}$)	k_2/k_2*
—	—	0.623	1.0
NaCl	3×10^{-1}	0.743	1.2
DACS	3×10^{-3}	2.69	4.3
MBz	3×10^{-3}	7.72	12
MNBz	1.5×10^{-3}	18.7	30

25 °C; the macroions are copolymers of the structural formula [5]
with $R_1 = R_2 = CH_3$ for DACS, $R_1 = CH_3$, $R_2 =$ benzyl for MBz,
and $R_1 = CH_3$, $R_2 = p$-nitrobenzyl for MNBz.

small deceleration. On the other hand, hydrophobic anionic macroion (NaPSS) clearly decelerates the reaction. As is shown in Table 3.7, the hydrophobic cationic macroions are expected to show hydrophobic attraction and, at the same time, electrostatic attraction for the dye cations. The fact that the reaction was accelerated indicates that the latter was overwhelmed by the former. Between the hydrophobic anionic macroion and the dye cation, there may exist simultaneously Coulombic and hydrophobic interactions. On the other hand, $OH^{\ominus}$ is repelled by PSS anions as a consequence of Coulombic repulsion; thus the two reactants, dye cations and $OH^{\ominus}$, are strongly separated from each other.

Table 3.7. *Interaction between macroions and reactants in the alkaline fading reaction of triphenyl methane dyes*

Macroions	$OH^{\ominus}$	Dye cation	Effect
Hydrophobic macrocation	Coulombic attraction	Hydrophobic attraction $\gg$ Coulombic repulsion	Acceleration
Hydrophilic macrocation and macroanion	Coulombic repulsion	Coulombic repulsion	Small deceleration
Hydrophobic macroanion	Coulombic repulsion	Hydrophobic attraction and Coulombic attraction	Deceleration

Fig. 3.10. Influence of various macroions on the alkaline fading reaction of ethyl violet (EV). $[EV] = 1.05 \times 10^{-5}$ M, $[OH^{\ominus}] = 1.05 \times 10^{-2}$ M; 30 °C. See text for the abbreviations of macroions. (T. Okubo & N. Ise, *J. Amer. Chem. Soc.*, **95**, 2293 (1973).)

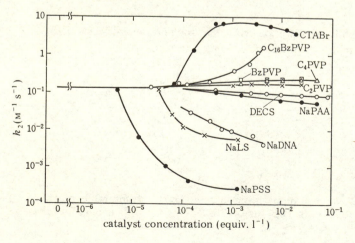

Also in Fig. 3.10, the influence of cetyltrimethylammonium bromide (CTABr), sodium lauryl sulphate (NaLS), and the sodium salt of DNA (NaDNA) are shown. CTABr, which is hydrophobic and cationic, shows greater acceleration than $C_{16}BzPVP$, and anionic NaLS has a retarding influence as does NaPSS. The deceleration demonstrated by NaDNA indicates the anionic and hydrophobic character of DNA under the experimental conditions.

The quantitative analyses of the activation parameters become intricate when both electrostatic and hydrophobic interactions are responsible for the observed catalyses. Table 3.8 gives the activation parameters of the fading reaction of ethyl violet catalysed by a cationic macroion and an anionic one. In the present case, equation (3.15) must be modified to equation (3.19) in order to include the volume change ($\Delta V_3^{\neq}$) of the iceberg structure around hydrophobic solutes.

$$\Delta V^{\neq} = \Delta V_1^{\neq} + \Delta V_2^{\neq} + \Delta V_3^{\neq} \tag{3.19}$$

For the fading reaction, $\Delta V_3^{\neq}$ may be expected to be negative, because the neutral activated complex ($Dye^{\neq}$) is supposed to be surrounded by larger iceberg structures than that of the cationic dye ($Dye^{\oplus}$). The underlying idea is that the electrical charge on solute species destroys, at least partly, the iceberg structure around the hydrophobic moiety of solutes (reactants and activated complex). It is also known that the contribution of the hydrophobic hydration to the partial molar volume of hydrocarbons such as methane, propane and benzene is negative. For example, the partial molar volumes of the aliphatic and aromatic compounds are smaller in water than those in pure liquids, by *c.* 20 ml mol^{-1} and 6 ml mol^{-1}, respectively. Thus $\Delta V_3^{\neq}$ is negative. On the other hand, $\Delta V_2^{\neq}$ may be reasonably expected to be positive for the reaction between cationic and anionic reactants because water molecules are liberated from the hydration shell in the course of activation. Furthermore, in the fading reaction, $OH^{\ominus}$ attacks the central carbon atom distant from the cationic nitrogen. Thus, the

Table 3.8. *Activation parameters for the alkaline-fading reaction of ethyl violet at 25 °C and 1 atmosphere*

Poly-electrolyte	Concentration (equiv l^{-1})	$\Delta G^{\neq}$ (kJ mol^{-1})	$\Delta H^{\neq}$ (kJ mol^{-1})	$\Delta S^{\neq}$ (J K^{-1} mol^{-1})	$\Delta V^{\neq}$ (ml mol^{-1}
None	0	79.4 ± 0.4	57.3 ± 2.1	-75 ± 8	-2.1 ± 1
$C_{16}BzPVP$	7.7×10^{-5}	79.0	49.3	-96	-3.2
	7.7×10^{-4}	75.2	25.0	-170	-4.4
NaPSS	1.2×10^{-6}	79.8	61.0	-63	-1.7
	1.2×10^{-5}	86.1	151	$+220$	$+8.5$

[Ethyl violet] $= 5 \times 10^{-6}$ M, [NaOH] $= 2 \times 10^{-2}$ M.

electrostatic hydration effects may be assumed to be not important. In other words, $\Delta V_2^{\neq}$ is positive but close to zero. Thus, the observed negative $\Delta V^{\neq}$ indicates the predominant role of hydrophobic hydration over electrostatic hydration, since $\Delta V_1^{\neq}$ is known to be negative or near zero.

That the larger iceberg structure was stabilized around the activated complex rather than around the reactants should imply a lowering of the disorderliness in the course of activation. This is consistent with the observed negative $\Delta S^{\neq}$; the entropy contribution from the electrostatic hydration, which should result in a positive change (see $\Delta S^{\neq *}$ for reaction (3.11) in Table 3.5), is negligible in the present case for the reason stated above.

When $C_{16}BzPVP$ is added to the fading reaction, the cationic macroion attracts, in addition to $OH^{\ominus}$, the neutral activated complex $(Dye^{\neq})$ more strongly than the cationic dye $(Dye^{\oplus})$. Therefore the iceberg structure of water should be incorporated into the bulky iceberg to a larger extent than the dye cations, so that the activated complex and the cationic macroions should be incorporated into the bulky, ice-structure to a larger extent than the reactant, causing a decrease in the partial molar volume (see Fig. 3.11). In other words, $\Delta V_3^{\neq}$ decreases. A decrease in $\Delta V_2^{\neq}$ with polymer addition is also reasonable, and $\Delta V_1^{\neq}$ is an intrinsic quantity, independent of the macroions. Thus $\Delta V^{\neq}$ decreases when $C_{16}BzPVP$ is added.

In the case of NaPSS, a hydrophobic anionic polyelectrolyte, the attraction toward the dye cation $(Dye^{\oplus})$ by the macroanion is much stronger than toward the activated complex. In other words, $Dye^{\oplus}$ will be found in the very vicinity of macroions with a greater probability than $Dye^{\neq}$. Thus, we expect more enhanced incorporation of $Dye^{\oplus}$ into the iceberg structure around the macroion than $Dye^{\neq}$ (Fig. 3.11), which should result in an increase in $\Delta V_3^{\neq}$ with polymer addition. Since $\Delta V_2^{\neq}$ is expected to decrease only slightly, as was stated above, $\Delta V^{\neq}$ should increase. This is exactly what is shown in Table 3.8.

3.4 'Catalysis' by macroions and the role of solvent

In the foregoing sections, it has been discussed how macroions accelerate or retard interionic chemical reactions. The studies have been carried out in aqueous media. The remarkable catalyses demonstrated in such systems are related to the following three factors:

(1) The high dielectric constant of water facilitates ionic dissociation of substrate and polyelectrolyte. Therefore, the substrate can interact with macroions electrostatically so that chemical reactions can be affected strikingly.

(2) Ionic reactants and/or the activated complex strongly interact with – or are hydrated by – water molecules which have a high dipole moment. This interaction is disturbed by the electrostatic field of macroions; dehydration takes place to cause catalysis.

(3) Strong interaction between water molecules – hydrogen bond formation – results in iceberg formation and hence hydrophobic attraction between hydrophobic solutes.

In order to confirm further these functions, the use of non-aqueous solvents is profitable. Especially if the role of solvation (or hydration) is to be correctly evaluated, organic solvents, which can preferentially solvate by solute species, are desirable. The following examples illustrate clearly what kind of role the solvent plays in macromolecular catalysis.

Fig. 3.11. Schematic representation of the change of the iceberg structure in the alkaline fading reaction in the absence (a) or presence of hydrophobic macrocations (b), or hydrophobic macroanions (c). Each V-shaped symbol indicates a water molecule; its ordered arrangement reflects the iceberg structure.

(a) Without polymer :

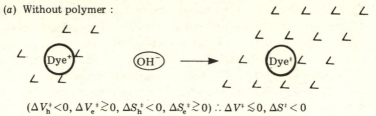

$$(\Delta V_h^{\ddagger} < 0, \Delta V_e^{\ddagger} \gtrsim 0, \Delta S_h^{\ddagger} < 0, \Delta S_e^{\ddagger} \gtrsim 0) \therefore \Delta V^{\ddagger} \lesssim 0, \Delta S^{\ddagger} < 0$$

(b) With hydrophobic cationic polymer:

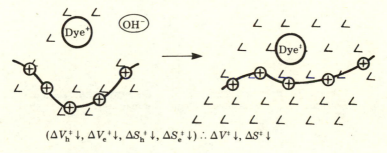

$$(\Delta V_h^{\ddagger} \downarrow, \Delta V_e^{\ddagger} \downarrow, \Delta S_h^{\ddagger} \downarrow, \Delta S_e^{\ddagger} \downarrow) \therefore \Delta V^{\ddagger} \downarrow, \Delta S^{\ddagger} \downarrow$$

(c) With hydrophobic anionic polymer:

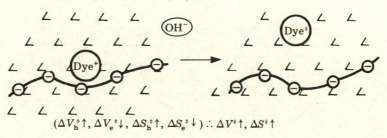

$$(\Delta V_h^{\ddagger} \uparrow, \Delta V_e^{\ddagger} \downarrow, \Delta S_h^{\ddagger} \uparrow, \Delta S_e^{\ddagger} \downarrow) \therefore \Delta V^{\ddagger} \uparrow, \Delta S^{\ddagger} \uparrow$$

The cyanoethylation of amino acids (reaction (3.20)) proceeds in water. It is accelerated when dimethylsulphoxide (DMSO) is added to the system; e.g.

$$\overset{\text{H}}{^{\ominus}\text{OOC}-\underset{\text{R}}{\text{C}}-\text{NH}_2} + \text{CH}_2=\text{CHCN} \rightarrow \overset{\text{H}}{^{\ominus}\text{OOC}-\underset{\text{R}}{\text{C}}-\text{NHCH}_2\text{CH}_2\text{CN}} \qquad (3.20)$$

the second-order rate constant k_2^* in water at pH 8.48 is $4.8 \times 10^{-4}\,\text{M}^{-1}\,\text{s}^{-1}$, and in a mixture of DMSO–water (50% v/v) it is $5 \times 10^{-3}\,\text{M}^{-1}\,\text{s}^{-1}$. This acceleration is attributed to enhanced nucleophilicity of the amino acid, which was caused by binding of the DMSO molecule (as a hydrogen-bond acceptor) toward the acid (see Fig. 3.12). On the other hand, a water molecule can act as both an acceptor and a donor of the hydrogen bond, so that no change in the nucleophilicity is observable. When cationic macroions, which are believed to be solvated strongly by DMSO, are added into the binary mixture, the nucleophilicity, once enhanced by DMSO, is expected to be lowered. As is shown in Fig. 3.13, for the L-phenylalanine–acrylonitrile reaction, addition of polyvinylpyridine quaternized by benzyl chloride (BzPVP) in the DMSO–water mixture (curves 2 and 4) gives smaller k_2/k_2^* values than in pure water (curves 1 and 3). Especially at pH $= 8.48$, k_2/k_2^* in the DMSO–water mixture is smaller than unity. This implies that BzPVP cations deprive DMSO molecules of the amino acid molecules so that the nucleophilicity of the amino acid reverts to the normal state which is found in pure water.

Similar solvent effects are found also for ester hydrolyses. The alkaline hydrolysis of *p*-nitrophenyl acetate was studied in the binary mixture of hexanol

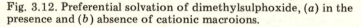

$$\qquad (3.21)$$

PNPA

Fig. 3.12. Preferential solvation of dimethylsulphoxide, (*a*) in the presence and (*b*) absence of cationic macroions.

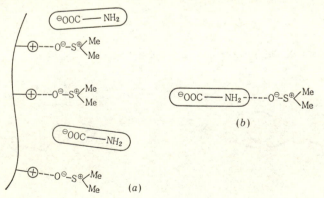

and water. The influence of polyethylenimines partially quaternized by alkyl (R) bromide (C_8PEI, C_{16}PEI, [8]) is shown in Fig. 3.14. Lowering of the water content enhanced the reaction at lower polymer concentration, though graphical presentation of the data is omitted. Generally speaking, the rates of ion–molecule reactions are dependent on the dielectric constant of the solvent.

Fig. 3.13. Influence of cationic polyelectrolyte on the cyanoethylation of L-phenylalanine at 30 °C. [acrylonitrile] = 0.2 M [L-Phe] = 10^{-3} M, curve 1: pH 10.06, 2: DMSO 50%, pH 10.06, 3: pH 8.48, 4: DMSO 50%, pH 8.48. (K. Yamashita, H. Kitano & N. Ise, *Macromolecules*, **12**, 34 (1979).)

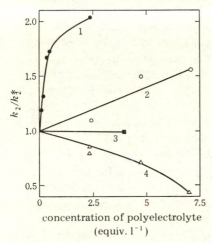

concentration of polyelectrolyte
(equiv. l^{-1})

Fig. 3.14. The alkaline hydrolysis of *p*-nitrophenylacetate in the binary mixture of *n*-hexanol and water, and the influence of polyethylenimine derivatives. ($\triangle$): C_{16}PEI, ($\bigcirc$): C_8PEI, [PNPA] = 2.5×10^{-5} M, [OH$^\ominus$] = 10^{-3} M, [H_2O] = 5.56×10^{-2} M. (T. Ishiwatari, T. Okubo & N. Ise, *Macromolecules*, **13**, 53 (1980).)

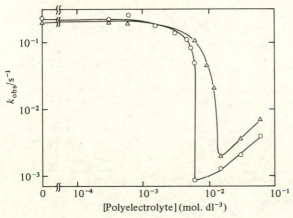
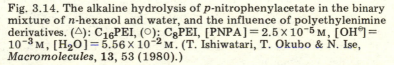

[Polyelectrolyte] (mol. dl^{-3})

$$-CH_2-CH-\overset{\cdot\cdot}{\underset{\underset{Br^{\ominus}}{|}}{N}}\overset{\oplus}{}-CH_2-CH_2-\overset{H}{\underset{\underset{R}{|}}{N}}\overset{\oplus}{}-CH_2-CH_2-\overset{H}{\underset{}{N}}-$$

$$Br^{\ominus}$$

[8]

Because the estimate of the volume of the activated complex is rather difficult, it is not obvious whether the observed increase in the rate is due to the change in dielectric constant. One of the simplest interpretations suggests that the $OH^{\ominus}$ is less hydrated at lower water concentration so that its reactivity is raised. When the concentration of R–PEI is increased, the rate drops quite sharply. This may be taken as implying that $OH^{\ominus}$ is concentrated around the cationic macroions by electrostatic interaction and, at the same time, water is condensed in the same region as a consequence of the hydrophilicity of the macroions. Thus $OH^{\ominus}$ in the vicinity of the macroions is hydrated, so that it loses the high activity which it would have had in the bulk phase of the binary solvent. The linear increase of the rate constant above $[R-PEI] = 10^{-2}$ equiv l^{-1} is due to the nucleophilic attack of free nitrogen in R–PEI. This interpretation is supported by the agreement of the rate constant observed in the presence of $OH^{\ominus}$ with that obtained without adding $OH^{\ominus}$.

The two cases discussed above indicate that the redistribution of solvent molecules is important to the observed polymer catalysis. In other words, we have a non-homogeneous distribution of solvent when macroions are added to multicomponent solvent systems. This kind of non-homogeneity is conceivable when specific interaction exists between solvent and macromolecules. However, this does not rule out the possibility of non-specific action of solvent in macromolecular catalysis. For example, the hydrolysis of hydrophobic ester is less strongly catalysed by likewise hydrophobic macroion in acetone-rich aqueous media than in acetone-poor environment, as is shown in Table 3.9. This result was attributed to weakening of the hydrophobic interaction in an acetone-rich environment.

Table 3.9. *Hydrolyses of n-butyl acetate and the influences of polymeric sulphonic acid and acetone*

	k_2/k_2*	
Acetone (% v/v)	PS–S(40)	PSS
0	6.3	1.9
15	2.6	1.5
30	1.6	1.1

PS–S(40): 40%-sulphonated polystyrene; PSS: polystyrene sulphonic acid.
[ester] $= 2.85 \times 10^{-2}$ M, [catalyst] $= 5 \times 10^{-3}$ M; 40 °C.

3.5 Catalysis by other interactions

Electrostatic interaction is of a long-range nature and causes remarkable catalysis on interionic reactions. However, it is unlikely that the relative orientation of the reacting partners can be controlled by this interaction. Hydrophobic interaction forces hydrocarbon solutes to get close to each other, but the regulation of spatial orientation of the solutes using this interaction only is not easy, although not impossible. Consequently, stereochemical effects in catalysis would not be likely if only electrostatic and hydrophobic interactions could be expected between the macromolecular catalyst and the reactants.

In the case of hydrogen-bond formation the situation is different. Because of the short-range nature, and also because this interaction is possible only between a pair of specified groups, it may be anticipated that introduction of these groups into the substrate and catalyst results in some sort of stereochemical regulation. In the following, several cases will be described, in which hydrogen-bond formation appears to play an important role.

3.5.1 *Asymmetric catalyses by polypeptides*

Ring-opening polymerization of the *N*-carboxy anhydrides (NCA) of α-amino acids takes place by attack of primary amines. When this type of reaction occurs on terminal amino groups of growing poly(amino acids), the end product is compounds of high molecular weight. It has been found by Bamford and his associates that this reaction proceeds faster with poly(amino acids) than with low molecular weight amines. The interpretation is that NCA molecules are concentrated along the poly(amino acid) by hydrogen-bonding. It is reported that the optical rotation of the product polymer changes with time when NCA containing either of the enantiomers in excess is used as the starting material. This time-change is possible only when preferential selection of the enantiomers takes place in the polymerization process. The most remarkable cases are alanine NCA and γ-benzyl glutamate NCA. Their polymers are considered to take on the α-helical structure and the orientation of an NCA

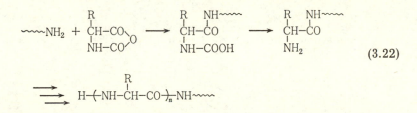

$$(3.22)$$

molecule is suggested to be regulated as a result of hydrogen-bond formation between the CO group of NCA and the NH group close to the terminal of the α-helix, as shown in Fig. 3.15.

Similar, but more striking stereochemical selection has recently been found by Inoue *et al.* for the following reaction.

$$CH_3(CH_2)_{11}SH \; + \; \underset{\underset{O}{\overset{\|}{}}}{CH_2=\overset{\overset{CH_3}{|}}{C}-C-CH_3} \; \longrightarrow \; CH_3(CH_2)_{11}S-CH_2-\overset{\overset{CH_3}{|*}}{\underset{\underset{O}{\overset{\|}{}}}{CH}-C-CH_3} \qquad (3.23)$$

The optical yield of the reaction becomes larger when poly(γ-benzyl-(S)-glutamate) (PBLG$_n$ [9]; n, degree of polymerization) or poly((S)-alanine) (PLA$_n$) [9] is added, as shown in Table 3.10. It is worth mentioning that the optical yields in the polymer-containing systems are higher than those in low molecular weight amine systems. The largest value is 47% in the PBLG$_{10}$-dichloroethane/ethanol system. It is proposed that the carbonyl group of the ion pair-type intermediate [10] forms a hydrogen bond with the NH groups

Table 3.10. *Influence of poly(α-amino acid) on the asymmetric addition of 1-dodecanethiol onto methylisopropenylketone*

Catalyst	Solvent	$[\alpha]_D$ of the product	Optical yield (%)
(S)-1-methylpropylamine	—	−0.02	0.1
Poly((S)-isobutyl ethylenimine)	Dioxane	+0.96	6
PBLG$_{10}$	Chloroform/ethanol	−6.4	37
PBLG$_{10}$	1,2-Dichloroethane/ ethanol	−8.2	47
PBLG$_{20}$	1,2-Dichloroethane/ ethanol	−7.2	41
(S)-alanine-N-propylamide	Chloroform/ethanol	−1.06	6
PLA$_3$	Chloroform/ethanol	−2.96	17
PLA$_5$	Chloroform/ethanol	−2.14	12

Fig. 3.15. Stereochemical control by α-helical structure in the polymerization of NCA. (H. Weingarten, *J. Amer. Chem. Soc.*, **80**, 352 (1958).)

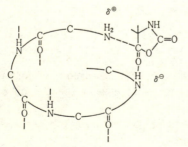

$$(CO-NH-CH)_n$$
$$|$$
$$R_1 \qquad R_1 = (CH_2)_2COOCH_2C_6H_5 \quad \text{for } PBLG_n$$
$$= -CH_3 \qquad \text{for } PLA_n$$

[9]

$$CH_3$$
$$|$$
$$CH_3(CH_2)_{11}S-CH_2-C^{\ominus} \qquad H_3N^{\oplus}\!\!\sim\!\!\sim\!\!R, \quad NH_2\!\!\sim\!\!\sim\!\!R: \text{poly}(\alpha\text{-amino}$$
$$|$$
$$COCH_3 \qquad\qquad\qquad\qquad\qquad\qquad\qquad \text{acid)}$$

[10]

near the end of the poly(amino acid) in a similar manner to that shown in
Fig. 3.15 and the steric hindrance effect between the intermediate and the end
group regulates the orientation of the carbanion and proton addition to the
carbanion sterically.

Optical yields as high as 90% are obtained in the asymmetric addition of
hydrogen cyanide to benzaldehyde, a cyanohydrin synthesis, catalysed by
a cyclic dipeptide, cyclo((S)-Phe-(S)-His), in which the imidazolyl group of
the histidine residue is catalytically active as a base.

$$C_6H_5-CH=O + HCN \rightarrow C_6H_5-\overset{*}{C}H-CN \qquad\qquad (3.24)$$
$$|$$
$$OH$$

In clear contrast, the corresponding linear dipeptide, Z-(S)-Phe-(S)-His-OMe
(Z: $C_6H_5CH_2OCO-$), exhibits almost no specificity (0.6% optical yield), indicating that the rigid conformation of cyclic dipeptide is essential for this
asymmetric synthesis.

3.5.2 *Template polymerization by macromolecules containing nucleic acid bases*

An example, in which marked specific interaction or orientational
control is found between two molecules by hodrogen-bond formation, is the
complementary interaction between polynucleotides. According to Watson
& Crick, two or three hydrogen bonds are formed between thymine and adenine
or between cytosine and guanine, respectively. These hydrogen bonds determine the distance between the two nucleic acid bases and force the bases to
locate in the same plane. The preciseness of the complementary interaction is
understandable in light of the fact that hereditary information is transmitted
quite exactly from one generation to another. Experiments have been reported
on macromolecular catalyses that make use of hydrogen-bonding. One example
is the polymerization of monomers containing nucleic acid bases in the presence
of macromolecular catalysts that also contain nucleic acid bases. For example,
the presence of a macromolecular catalyst obtained from a monomer containing
thymine (MAOT, [11]), enhances the free-radical polymerization of a monomer

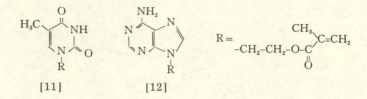

[11]   [12]

(MAOA, [12]) having an adenine moiety, as shown in Fig. 3.16. On the other hand, the polymerization of MAOT under the same conditions is slightly hindered with increasing polymer concentration. These facts are accounted for in terms of the hydrogen-bond formation between complementary MAOT and MAOA.

The polymerization of mononucleotide has been carried out in the presence of oligonucleotide bound on the surface of crosslinked macromolecules. The complementary oligonucleotide (degree of polymerization 3) is synthesized only when the template oligonucleotide has a degree of polymerization larger than 4.

The concept 'template polymerization' was first introduced into the chemistry of synthetic macromolecues by Bamford for the free-radical polymerization of acrylic acid, enhanced by the addition of polyethylenimine. The interpretation for this acceleration caused by the latter is that the monomers become absorbed on to template macromolecules. The acceleration, however, cannot be observed if the low molecular weight analogue, tetraethylene pentamine, is used in the place of the polyethylenimine. Here again, the advantage of macromolecules over low molecular weight substances is clearly demonstrated (cf. Fig. 3.3).

Fig. 3.16. The influence of macromolecules on the polymerization of monomers containing nucleic acid bases. Solvent: pyridine. 60 °C; 3 h. [monomer] $= 2 \times 10^{-2}$ M [radical initiator] $= 10^{-3}$ M. MAOZ: MAOA (●) or MAOT (○). (K. Takemoto, Y. Inaki, Kagaku-sohsetsu (chemistry review) No. 17, '*Interaction of Macromolecules and their Functions*' Tokyo University Press (1977).)

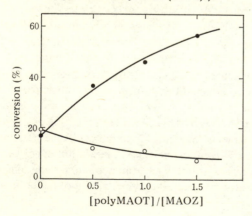

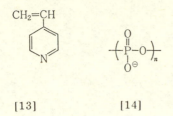

[13]   [14]

A similar tendency is observed for the polymerization of 4-vinylpyridine [13] in the presence of polyphosphate [14], polystyrene sulphonate [3], or poly-ethylene sulphonate [4]. Although it is not established that the monomer molecules are linked up along the template macromolecules and 'zipper-like' polymerization takes place, as is often schematically illustrated, the acceleration by the template is obvious. Interestingly enough, polyvinylpyridine is reported to have a stereospecific structure when prepared in the presence of polyphosphate.

3.5.3 *Catalyses by polymer–metal complexes*

Many metal ions are known to form metal complexes with ligands. There exist a variety of biopolymer–metal complexes, such as the metallo-enzymes. With respect to their remarkable functions, keen attention is recently being paid to complexes of metals with synthetic macromolecules. We discuss here recent studies on the catalytic function of the polymer–metal complexes; the methods of synthesis, the structure and most properties of the complexes will not be treated.

Poly(vinyl alcohol) [15] forms complexes with Cu(II) or Fe(III). The Cu complex has a planar structure ($Cu(O)_2(OH)_2$), whereas the Fe complex has an

$$-CH_2-CH-CH_2-CH-$$
$$OH \qquad OH$$

[15]

octahedral structure ($Fe(O)_3(OH)_3$). They differ also in stability: the stability constant of the Fe complex is 10^{10}-times larger than that of the Cu complex. It is also possible to prepare a mixed complex with Cu and Fe in a single polymer chain. As shown in Fig. 3.17, the mixed complex is more efficient than the Fe complex in the decomposition of hydrogen peroxide. Because of the high stability of the Fe(III) complex, it is suggested, the conformation of the polymer chain is modified in such a way that the original coordination structure of Cu(II) is no longer stable and hence binding of the reactant becomes easier.

Phenol derivatives are known to polymerize by the catalytic action of metal complexes in the presence of oxygen. For 2,6-dimethylphenol, the polymerization is suggested to involve the four elementary steps shown. The substrate anions

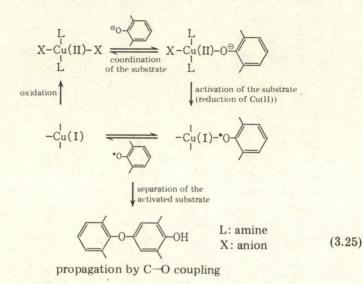

propagation by C–O coupling

coordinate toward the complex, the Cu(II) is reduced (or the substrate is activated), and then the activated substrate separates, and grows to polymers by C-O coupling. The Cu(II) complex is regenerated by oxidation of the Cu(I) complex left. In many cases, polymeric ligands are found to accelerate the polymerization, though only slightly. Table 3.11 shows some examples. Clearly, the polymeric complex (QPVP-Cu) has a higher efficiency than low molecular weight complex (Py-Cu) in dimethylsulphoxide. As a cause of this acceleration,

Fig. 3.17. Decomposition of hydrogen peroxide by a mixed complex of Cu(II) and Fe(III) with poly(vinyl alcohol) (PVA). [complex] = 2×10^{-5} M, $[H_2O_2] = 4 \times 10^{-2}$ M, ionic strength = 0.1 [KCl]; pH 10.5; 25 °C. Curve a: Fe(III)-PVA, Curve b: mixed complex. (H. Shirai *et al.*, *J. Chem. Soc., Japan*, **1978**, 117.)

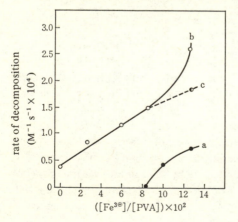

it is suggested, the activation of the substrate becomes easier when the polymeric complex is used; the coordination bond of Cu is under stress because the polymer chain is forced to fold to form the complex. As a matter of fact, the activation energy for this process is found to be about $30\,kJ\,mol^{-1}$ for the polymeric complex whereas it is $80\,kJ\,mol^{-1}$ for the low molecular weight counterpart. The influence of the polymeric complex depends on the kind of solvent. For example, PVP-Cu is as active as Py-Cu when benzene is used as the solvent. This would be due to the conformation of the polymer chain varying with the solvent used.

3.6   Heterogeneous macromolecular catalysis

The macromolecular catalyses discussed above have been studied in homogeneous systems, in which both reactants and catalysts are dissolved in solvent (in most cases water) homogeneously. The study of such systems can provide useful, basic information on the mechanism of catalysis, but heterogeneous catalysts appear to be more practical, since in this case separation of the products from the catalysts is easy.

The first attempt reported was with ion-exchange resins which are crosslinked, insoluble polyelectrolytes. In the early 1950s, ester hydrolysis was studied in a column of sulphonated polystyrene beads and the catalyst efficiency was found to be hardly affected by the velocity with which the reactant solution passed through the column. The dimension of the beads seems to play an important role; the diffusion of the solute into the beads affects greatly the catalytic efficiency. Thus crosslinked macromolecular catalysts are not always more advantageous than the low molecular weight ones. For example, the ratio of the catalytic efficiencies of a cationic exchange resin and hydrochloric acid is 0.5, 0.3 and 0.05 for hydrolysis of methyl acetate, ethyl acetate, and ethyl-*n*-caproate, respectively. On the other hand, the investigation on esterification, alcoholysis, acetal condensation and inversion of sugar indicates in most cases

Table 3.11. *Influence of Cu complexes on the oxidative polymerization of 2,6-dimethylphenol*

Catalyst	Solvent	[Cu] (M)	Reaction time (h)	Polymer yield (wt%)	Mol. wt of polymer ($\times 10^{-3}$)
QPVP-Cu	DMSO	0.01	1	96	1.5
Py-Cu	DMSO	0.01	5.5	64	1.6
PVP-Cu	Benzene	0.005	1.5	99	7.1
Py-Cu	Benzene	0.005	1.5	88	5.8

QPVP: partially quaternized polyvinylpyridine, Py: pyridine, PVP: polyvinylpyridine, DMSO: dimethylsulphoxide.

that heterogeneous macromolecular catalysts are more efficient than corresponding low molecular weight ones.

The second example is the fixation of metal complexes on a macromolecular matrix. For example, the Wilkinson complex [16] is an efficient catalyst for hydrogenation of olefins. When the complex is bound on polystyrene resin (polystyrene crosslinked with divinyl benzene), it can also be used as an excellent catalyst [17]. One advantage is the stability: the low molecular weight complex

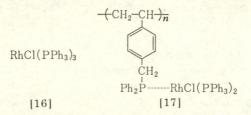

RhCl(PPh₃)₃

[16]   [17]

loses its activity when exposed to air, whereas the polymer-bound complex is stable. This is probably due to Rh atoms being buried in the deep interior of the resin. In this respect, it must be mentioned that [17] may be less active than [16] depending on the dimension of the substrate. As seen from Table 3.12, the resin-type catalyst [17] shows a more enhanced substrate specificity than the low molecular weight complex [16].

Polymer complexes are known to affect not only the reaction rate but also the route of the reaction. For example, the hydroformylation of 1-pentene (3.25) gives a value of 3 for n/i, where n and i are the quantities of unbranched and branched aldehydes, when use is made of a low molecular weight complex [18] whereas a polymeric complex [19] has $n/i > 12$ under the same conditions.

$$H_3C{\overset{CH_2}{\diagdown}}CH_2\text{–}CH{=}CH_2 \xrightarrow{H_2:\ CO} CH_3\text{–}CH_2\text{–}CH_2\text{–}CH_2\text{–}CH_2\text{–}CHO$$

$$\begin{array}{c} CHO \\ | \\ + \ \ CH_3\text{–}CH_2\text{–}CH_2\text{–}CH\text{–}CH_3 \end{array} \qquad (3.26)$$

Table 3.12. *Relative rate of hydrogenation of olefins*

Olefin	Polymeric complex [17]	Low mol. wt complex [16]
1-Hexene	2.55	1.4
Cyclohexene	1	1
Cyclododecene	0.22	0.67
Δ²-cholestane	0.03	0.71

In benzene; 25 °C. Hydrogen pressure 1 atm ($=1.01325 \times 10^5\,Nm^{-2}$). The rate is relative to cyclohexene.

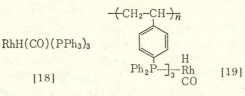

RhH(CO)(PPh$_3$)$_3$

[18]

[19]

Recently, intensive research is being carried out on chemical and physical immobilization of enzymes. These heterogeneous enzymes systems allow us to carry out complicated reactions with a high efficiency and repeatedly without losing expensive enzymes.

B POLYMER CATALYSTS WITH ENZYME-LIKE ACTIVITIES
IWAO TABUSHI

3.7 Polymer catalysts having functions like enzymes

Polymer catalysts with enzyme-like activities should:

1. *Have a large rate acceleration.* A polymer catalyst may be used even under mild conditions such as normal temperature, normal pressure, and neutral pH in dilute aqueous solution.
2. *Have efficient turnover or recycling.* Even a small amount of polymer catalyst should be effective.
3. *Be able to select the **right** substrate from a complicated mixture of compounds.* The polymer catalyst should be able to 'recognize' the *right* substance among coexisting impurities in crude raw material. If this can be accomplished, crude raw material may be used in chemical industries.
4. *Be able to select the **right** reaction from many possible reactions.* A polymer catalyst should catalyse only the desired reaction when several are possible.
5. *Regulate or control the reaction.* A polymer catalyst should accelerate the reaction when a large amount of the starting material is present, but retard it when the starting material is deficient and/or an excess amount of product has been formed.

These requirements, especially the last one, have not been satisfactorily clarified. Regulation may be classified into two categories: 'local regulation,' where local information is treated within its local site (mostly by regulatory enzymes); or 'central regulation' where local information is transferred to a central regulation site and processed together with information from other sites (mostly by DNA control). A wide variety of regulation types exist between these two extremes. These regulation mechanisms must be clarified chemically in order to be able to apply them to practical purposes in chemical industries.

Although satisfying the fifth requirement will need further study, polymer catalysts having the other four requirements may now be prepared relatively easily by carrying out the following general procedures: (i) design a single small molecule which has the desired function, such as catalysis or metal binding; (ii) synthesize this molecule; (iii) attach this molecule to an appropriate 'parent polymer' or a readily available polymer via chemical bonding or physical interaction (ion-pairing, adsorption, chelation).

An alternative method is to design a molecular assembly (such as an inclusion complex, a metal cluster, a metal complex, or a micelle) in which each participating molecule is designed to show only a part of the total entire function and the proper synergetic combination of all parts leads to the desired 'total function'. A very sophisticated polymer having all the necessary functions at the appropriate sites is easy to imagine but very difficult to prepare.

This chapter discusses three procedures for preparing a polymer with enzyme-like catalytic activity:

1. *Chemical modification of commonly used polymeric material.* Chloromethylpolystyrene, polyvinyl alcohol and some derivatives of polyacrylamide are suitable polymeric materials. Appropriate functional (catalytic) groups can be introduced by simple chemical reactions.

2. *Polymerization of a monomer bearing suitable functional group(s).* A special vinyl monomer bearing a suitable catalytic group or bifunctional monomers bearing a third functional (catalytic) group may be used for vinyl polymerization or polycondensation, respectively. Special attention must be paid to 'ordered polymerization' when two or more functional groups are required in the catalytic site. Merrifield's method of preparing a polypeptide in the 'required order' is a very good possibility for polypeptides. Polymeric nucleotides and polymeric sugars of the required order may be similarly prepared.

3. *Attachment of an enzyme to a polymer or polymerization of the enzyme itself.* An immobilized enzyme is an example in which the basic characteristics (e.g. shape, physical properties) of the desired polymer are determined mostly by the nature of the parent polymer, and the catalytic activity is determined mostly by the enzyme. Immobilization may be conveniently achieved by (i) covalent bond formation, (ii) non-covalent bond (ion pair, metal coordination, adsorption, etc.) formation, or (iii) encapsulation into a matrix of a microcapsule.

In order to prepare a polymer catalyst with enzyme-like activity, detailed understanding of the nature of enzymes is important. For practical applications, experimental procedures are often limited to a narrow range in order to maintain the enzyme activity. For example, a polymer catalyst exhibiting only hydrolytic activity (without substrate specificity or regulation) can be prepared rather easily today. But higher sophistication is not yet possible and remains a future challenge.

3.8   Native enzymes and polymers

Many polymer chemists strongly believe, or feel, at least, that an appropriate polymer structure is the necessary and sufficient condition for obtaining the desired enzyme activity. However, this idea is simply based on the fact that all known native enzymes are polymers. It is now clear that the polymer structure is not a necessary condition for the preparation of an artificial enzyme, because many excellent artificial enzyme systems have been successfully prepared using micelles or inclusion complexes and no polymeric compound. Therefore, let us start our discussion with the relationship between enzyme activity and polymer structure.

Recent opinions of different authors on polymer catalysts with enzyme activities are summarized in Table 3.13. Despite superficial differences, all claim that an appropriately designed polymer may act as an enzyme-like catalyst. This concept is rephrased into the chemical expressions summarized in Table 3.14, which also shows typical examples of possible chemical approaches to mimic 'fragmental' (or a certain part of) enzyme activities using simple organic compounds.

If the readers accept the above common understanding, they will accept the following conclusive remark, too: direct utilization of polymeric material is not a necessary condition for the artificial construction of (a part of) the enzyme activity. This conclusion combined with the previous one would lead to the

Table 3.13. *Similarity between enzymes and polymer catalysts*

Book and authors	Similarities cited
Medical polymers[a] (Takemoto & Tabushi)	Fixed geometry of functional groups, suitable composition of reaction sites, flexibility, total function afforded by appropriate combination of local structures, long-range effect through polymer structure, intermolecular or intersubunit interaction
Enzymes and polymer catalysts (Narita, Imanishi, & Sakiyama) (in Japanese)	Interaction of functional groups, suitable composition of reaction sites, substrate binding, polymer-chain effects (conformational effect)
Polymer catalysts (Kunitake, Shimizu, *et al.*)[b] (in Japanese)	Suitable composition of active sites, cooperativity, conformational fit induced by substrate binding

[a] See Further Reading at end of book.
[b] These authors claim that enzyme-like catalysts should not be inorganic compounds of low molecular weight but organic compounds of high molecular weight.

following remark: appropriate utilization of polymeric materials may be a sufficient condition but cannot be a necessary condition for the artificial (chemical) construction of the enzyme activity; or, utilization of the polymer catalyst is one of the most appropriate tools for the chemical 'mimicking' (or modelling) of the enzyme activity.

How much can enzyme activities be approximated by polymer catalysts? Some enzyme activities seem to be too sophisticated and complex to be mimicked by simple polymer catalysts. These sophisticated functions are based on the unique and very precise 'information' which enzymes possess. This information is a kind of 'pattern information' related to the size and shape of the protein, which originates from DNA:

> enzyme function ⟵ shape, size or other properties of the enzyme as a whole (including various partial functions of enzymes) ⟵ higher dimensional structure of the protein (quarternary or tertiary structure) ⟵ amino acid sequence of the protein ⟵ base sequence of DNA

This type of information transfer is far beyond the present level of polymer chemistry. For example, if we compare the biological system and polymer chemistry related to polymerization of α-amino acids, polymer chemistry clearly cannot compete with the biological system. No procedure is yet known for the precise sequential polymerization of α-amino acids in polymer chemistry, except for Merrifield polymerization, which has difficulties, like contamination by undesired proteins and laborious procedures. Thus, industrial production of desired proteins in large quantities and at reasonable prices in short periods

Table 3.14. *Enzyme-like functions that can be realized by low molecular weight compounds*

Function	Possible replacement
Fixed geometry of functional groups	
Specific reaction 'field' (specific local reaction environment)	Inclusions, micelles
Conformational fit	Inclusions, crown ether, lipophilic interacting systems
Appropriate combination of local unit structure to afford total function	Functionalized micelles, functionalized inclusions
Remote interaction	(Not necessary?)

remains practically impossible. Although information transfer is commonly used in biological systems, the use of this procedure (the Merrifield type) does not seem appropriate for the preparation of polymer catalysts. At present, polymer catalysts of practical significance can best be prepared either by simple, classical polymerization procedures, like polyadditions and polycondensations, or by simple chemical modification of polymers. These simpler polymer catalysts possess a rather high 'randomness' of sequence which prevents sophistication of the artificial enzyme function that is acquired from precise sequential information.

Let us consider the active site of chymotrypsin. The structure and catalysis of this enzyme has been well investigated, and the key role in the catalysis is known to be played by the so-called 'charge-relay system' in which imidazole (His), hydroxyl (Ser) and carboxylate (Asp) are appropriately (linearly) aligned at a fixed distance of *c.* 0.3 nm apart. Copolymerization of a monomer bearing imidazole and a monomer bearing hydroxyl may provide this active site alignment but with rather low probability, since the polymer backbone can take various conformations and both imidazole and hydroxyl groups retain the freedom of internal rotation. In order to maintain the catalytic activity of a polymer at a certain level, the population of its catalytic functional groups should be reasonably large. This restriction does not allow further introduction of another functional group participating in the enzyme function since three-group contact must be much less probable. Thus, overall catalytic activity of the copolymer catalyst becomes much lower than that of native chymotrypsin due to the extremely low population of the active site in the polymer.

The situation becomes more serious when an attempt is made to mimic more sophisticated enzyme catalysis, which would require many (more than three) functional groups in precise spatial orientation to construct a catalytic site or a so-called 'recognition site'. The probability of obtaining this type of active site by simple polymerization of random orientation is not zero but is certainly negligible. In fact, no artificial polymer catalyst has ever 'recognized' or differentiated a slight change in an apolar (or hydrophobic) part of a substrate. In native enzymes, this precise *recognition* is achieved by their binding sites, so-called hydrophobic 'pockets' or 'clefts', each of which is very specific as to the shape of its substrate. This 'recognition site' or binding site is usually constructed of several hydrophobic amino acid residues which are tightly held by a covalent bond or hindered internal motion. The shape and the size of the pocket are therefore maintained with reasonable stability. Conveniently, the magnitude of the interaction between two amino acid residues is in the range of 10^0-10^1 kJ unit^{-1}, a bond which is readily cleaved and re-formed at room temperature. The so-called 'conformation change' of proteins, therefore, easily takes place, and the shape of the protein as a whole or the shape of the binding site changes without changing the amino acid sequence at room temperature. The function of the protein as a whole or the function of the active site also changes smoothly at room temperature. A binding site may change its shape in order to accommodate a specific guest. This change is often important for some successive phenomena such as 'induced fit', induced strain for acceleration, or *allosterism*.

These sophisticated functions of native enzymes originate from the information transferred from DNA or, in other words, from its low entropy character. According to the information theory, a long fixed sequence has a low probability and therefore is in a low entropy state, as is expressed by the following equation:

$$S = k \ln W$$

Synthetic polymers, after careful 'functionalization' or chemical modification (with high information or low entropy) may show such sophisticated functions. However, serious experimental problems remain in sequential polymerization and make sophisticated 'functionalization' extremely difficult. The introduction of several different functional groups in synthetic polymers at fixed positions, and in a fixed spatial orientation, is far beyond the conventional techniques of present-day polymer chemistry.

This problem of 'sophisticated functionalization' can be rather easily solved by the introduction of an active-site model or a coenzyme model which has been independently prepared. This will be discussed in the following section.

3.9   Introduction of prepared enzyme-like activity into polymers

As discussed in §3.8, the basic idea of using synthetic polymers as enzyme-like catalysts is not based on simple analogy of their molecular weights. The extremely sophisticated functions of native enzymes can only be 'prepared' by sacrificing a large entropy (negative entropy change) controlled by DNA information, while introduction of artificial enzyme function into the polymers is achieved by artificially implanting precise information. For example, precise information transfer can be achieved by precise sequential polymerization of a series of monomers.

Even though this procedure is not very practical, synthetic polymers still have some advantages as catalysts. First, they are versatile as convenient base materials. A polymer can be formed into a convenient shape (bead, plate, membrane, capsule, film, etc.) and after the catalytic reaction is over, it can be easily separated from water or other reaction media (as shown by a typical successful example of convenient polymer phase-transfer reagents). A second advantage is the conveniently attained appropriate 'functionalization' or introduction of appropriate functional groups to polymer materials. At present, very sophisticated multi-functionalization of a polymer (introduction of a series of necessary functional groups into a polymer) can be done by direct combination of a 'highly' functionalized catalyst like a pendant group with a simple base polymer, rather than attempting to introduce various functional groups at given sites in a polymer skeleton. Thus, we should try to prepare a compound with a sophisticated function of a low entropy structure. Let us consider this situation with two typical examples.

First, let us consider how to prepare a 'pocket' which specifically and strongly binds a few substrates, but binds other similar compounds very poorly.

3.9.1 *Specific binding of cupric ion*

Cupric ion binds ammonia to form the complex $Cu(NH_3)_4^{2\oplus}$ or with ethylenediamine to form $Cu(en)_2^{2\oplus}$, where the stability constant of the latter is much larger than the former. The difference in the binding constants of these two species is mostly due to differences in electron density and the entropy factor. Since two nitrogen atoms are fixed at the appropriate distance for *cis*-coordination in ethylenediamine, the entropy loss involved in the second coordination is much smaller (or the probability of the second coordination is much larger) compared with the corresponding coordination of NH_3. A similar *entropy factor* is also operative in the complexation with a crown ether compared with corresponding linear polyethers.

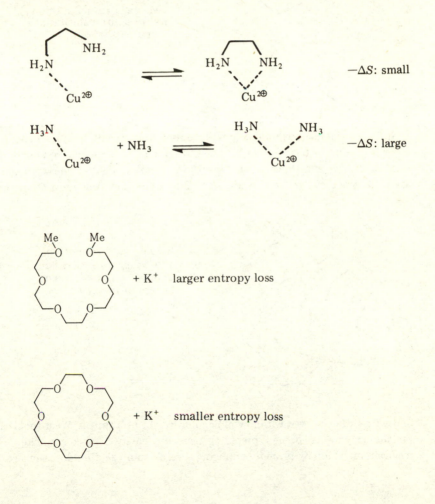

If this entropy factor plays a determining role in the complexation in general, a cyclic tetramine of appropriate ring size should form a $Cu^{2\oplus}$ complex (planar or tetrahedral, depending on its skeletal configuration) with a very large stability constant. This does happen and 'cyclam' or macrocyclic tetramine binds cupric ion 'specifically', or with an especially large stability constant.

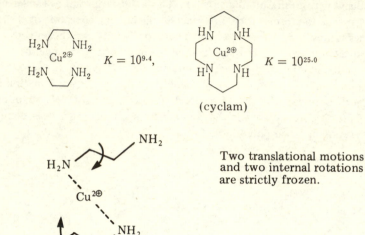

$$K = 10^{9.4}, \qquad K = 10^{25.0}$$

(cyclam)

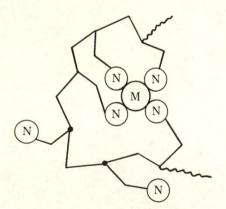

Two translational motions and two internal rotations are strictly frozen.

Alkali or alkaline earth metal ions are bound only weakly to *cyclam* and other transition metals, such as $Ni^{2\oplus}$ or $Co^{2\oplus}$, are bound much more weakly than $Cu^{2\oplus}$ to this cyclam. However, this specificity is not as high for linear polyamines as for the macrocycles. The specific $Cu^{2\oplus}$ binding is, therefore, most conveniently

Loss of internal motion and/or inappropriate coordination depends on the size and the charge of the metal ion.

and efficiently achieved with low molecular weight polyamines. What the native metallo-enzymes are doing is preparing a specific 'cavity' to accommodate a certain metal ion by properly arranging several amino acid residues such as

histidine, cysteine, aspartic acid, glutamic acid or lysine, as satisfactory participating ligands. The metal-binding ability thus attained is strong as shown for
carbonic anhydrase, carboxypeptidase, alkaline phosphatase, rubredoxin or
ferredoxin.

Next, let us compare two possible ways to obtain functionalized polymers of
strong metal-binding ability by taking the tetramine–$Cu^{2\oplus}$ complex formation as
an example:

- (*a*) An amino group is introduced into appropriate polymer skeletons with
  the expectation of statistical formation of the tetramine–$Cu^{2\oplus}$ complex.
- (*b*) Cyclam (macrocyclic tetramine) is introduced into the appropriate
  polymer skeletons.

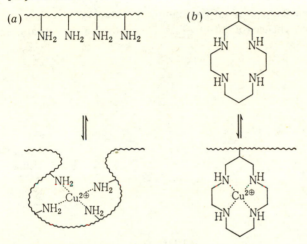

Polymers (*a*) may be prepared via many possible convenient pathways:

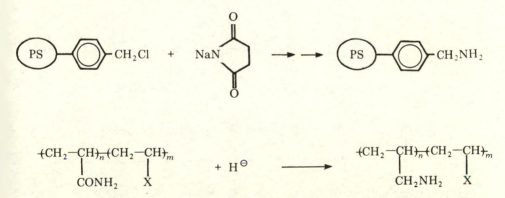

Preparation of polymers (*b*) requires preparation of macrocyclic polyamines in
advance. Preparation of polymers (*a*) seems much simpler, but the $Cu^{2\oplus}$ binding
ability is quite different, and cyclam attached to polystyrene binds $Cu^{2\oplus}$ much

more strongly than polymers (*a*). The $Cu^{2\oplus}$ binding by the cyclam polymer is stronger by several orders of magnitude than ethylenediamine attached to the polymer. The specificity of $Cu^{2\oplus}$ binding over $Ni^{2\oplus}$ or $Co^{2\oplus}$ binding is also much higher for the cyclam polymer. Table 3.15 shows the binding characteristics of the cyclam-polymer. Readers may notice that this is a typical example of the so-called 'macrocyclic entropy effect' generally seen in macrocyclic bindings. A large entropy loss usually associated with the complexation

$$M^{+m} + nL \rightleftharpoons M^{+m}Ln$$

depending on the number of bound ligand molecules, weakens the binding. But for a macrocyclic polydentate, loss of freedom of motion during the binding process is much smaller than the above complexation. The entropy loss associated with macrocycle ligation is therefore minimum, leading to strong binding, where a negative entropy change (or entropy loss) occurs during the preparation of a macrocycle itself.

Table 3.15a. *$Cu^{2\oplus}$ binding by cyclam bound to polystyrene (25 °C, pH 5.4)*

Polymer	Initial conc. of $Cu^{2\oplus}$ (ppm)	Equilibrated (final) conc. of $Cu^{2\oplus}$ (ppm)
(PS)—CH₂—N N ... (cyclam)	1270	2.5
(PS)—CH₂—N N ... (cyclam)	1270	3.8

Table 3.15b. *Relative binding of $Cu^{2\oplus}$ and $Ni^{2\oplus}$ by cyclam bound to polystyrene*

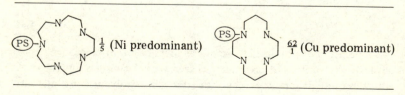

(PS)—N ... $\frac{1}{5}$ (Ni predominant)   (PS)—N N ... $\frac{62}{1}$ (Cu predominant)

3.9.2 Specific binding of phenols or benzoates

Let us proceed to the more sophisticated or complicated problem of 'pattern recognition' of molecular shape as a second example of preparing a compound with a sophisticated function. Let us take an aromatic compound bearing a negative charge as a model substrate to compare the binding characteristics between two types (a and b) of functionalized polymers.

Simple introduction of hydrophobic pendant groups (long alkyl chain) into polymers leads to self-aggregation of the pendant groups in water. This aggregation, called a hydrophobic domain or a 'polymer micelle', strongly binds various hydrophobic guest molecules. However, this binding is indiscriminate and all of the hydrophobic compounds are bound by the hydrophobic domain, just like micelles or emulsions. We may say that this indiscriminate hydrophobic binding is a rather poor function of the polymer. We are looking for a more sophisticated function to bind a hydrophobic molecule of a certain size and shape specifically. In fact, an enzyme usually 'recognizes' the shape and the size of the specific substrate. No polymer catalyst has yet been reported that shows the specific binding of a given substrate such as benzene.

This type of molecular recognition has been extensively studied in the organic and inorganic chemistry of lower molecular weight compounds. One of the best examples is cyclodextrins, which are a series of cyclic oligoglucoses including the hexamer, heptamer and octamer named in a conventional way, α-, β- and γ-cyclodextrin, respectively. Here again, these host molecules display the 'macrocyclic entropy effect', leading to the formation of a torus shape as a whole with a so-called hydrophobic cavity, which is of the low entropy state. Every glucose unit loses a considerable amount of freedom of motion and is oriented symmetrically by pointing three of the axial hydrogens toward the inside of the macrocycle cavity, two secondary hydroxyl groups toward the inside and outside from the bottom of the torus, and one primary hydroxyl toward the outside from the top of the torus. The cavity has a practically fixed size and is hydrophobic due to the surrounding C—H bonds. Cyclodextrins are soluble in water because $3n$ ($n = 6, 7, 8$) hydroxyls are oriented along the 'rim' of the torus. When a hydrophobic substrate, or a hydrophilic substrate bearing a hydrophobic substituent, is dissolved in water, it is strongly bound to the cyclodextrin cavity, if van der Waals contact of the substrate (guest) with the cavity wall is appropriate. And this van der Waals contact determines the host-guest specificity: for α-cyclodextrin, benzene and some substituted benzenes, cyclohexane and its simple derivatives; for β-cyclodextrin, naphthalene and its simple derivatives, adamantane and its simple derivatives; for γ-cyclodextrin, anthracene derivatives. This type of pattern recognition has seldom been observed for polymer catalysts.

Specific and strong recognition of an anion in water may occur due to complexation with a proper metal ion, as seen in the binding by metalloenzymes. This recognition combined with the hydrophobic shape recognition by the cyclodextrin cavity may offer some artificial hosts specifically recognizing a guest having both a negative charge and a hydrophobic substituent of a certain shape. These suggest an answer to the question of how to prepare a specific artificial host molecule for the binding of phenols or benzoates. Cyclodextrin having an arm to which a metal cation is attached is a promising candidate for the specific host.

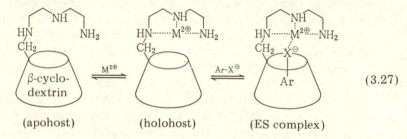

(apohost)   (holohost)   (ES complex)   (3.27)

This cyclodextrin flexibly capped with a metal ion has been prepared by the authors. The triamino group is a strong ligand toward certain metal ions such as $Ni^{2\oplus}$ and $Zn^{2\oplus}$, binding the metal ion nearly completely to form a host-metal complex quantitatively. We may call this complex a 'holohost' and the triamino cyclodextrin an 'apohost', after the conventional nomenclature of apoenzyme and holoenzyme. This holohost exhibits the 'double recognition' property where the cyclodextrin cavity recognizes a guest by its hydrophobic substituent of specific size and shape, and at the same time the metal ion recognizes a guest at its negative charge or its coordinating group in general. As a result, this holohost specifically binds a hydrophobic anion such as phenol, benzoate or adamantanecarboxylate. Table 3.16 shows typical examples from a series of measurements of holohost–substrate complexations. We are now considering a guest specifically bound to the holohost bears both a strong coordinating group toward $Zn^{2\oplus}$ and a strong hydrophobic moiety of a size that 'fits' (making van der Waals contact) the size of the cavity wall. 2-Adamantanone-1-carboxylate is the specific substrate which has the adamantane moiety for the ideal hydrophobic recognition and β-ketocarboxylate for the ideal complexation. The association constant of adamantanone-carboxylate toward the present holohost is extremely large, even more so than that of a specific substrate toward a common enzyme. Note that substrate specificity (ratio of the association or dissociation constant for a specific substrate to that for a certain nonspecific substrate) is also remarkable. The last column of Table 3.16 shows how the introduction of $Zn^{2\oplus}$ into the apohost enhances the recognition. We may estimate the elemental recognition free energy by using simplified model substrates. Interestingly, the 'total recognition free energy' is approximately given by the sum of each elemental 'recognition free energy' separately estimated as shown in Fig. 3.18. The presently observed additivity is very helpful for designing or 'building up' a hypothetical host molecule to be synthesized, which has the required total 'recognition free energy' (or association constant) toward a given guest molecule. Alternatively, we can design or 'build up' a hypothetical guest molecule to be synthesized, specific to a given enzyme. This 'building up' principle may be a helpful strategy for designing a molecule with a required function.

Can such complicated and sophisticated functions be obtained using a polymer? Unfortunately, at present, the answer is 'no', because of the infinitely small probability of preparing such a low entropy state as that shown in Fig. 3.18 by combining several statistically distributed and freely rotating hydrophobic pendant groups to form a considerably fixed cavity for the hydrophobic binding, and bringing two or three statistically distributed coordinating groups to form

Table 3.16. *Dissociation constants of typical hydrophobic anions. Triamino cyclodextrin* $\cdot Zn^{2\oplus}$ *complex*

Guest	Host	K_d (M)	K_d (β-CD) / K_d (holohost)
(adamantane-CO$_2^\ominus$)	β-CD	1.2×10^{-3}	
	Ia-Zn$^{2\oplus}$	3.6×10^{-6}	330
(adamantane-CO$_2^\ominus$)	β-CD	4.3×10^{-3}	
	Ia-Zn$^{2\oplus}$	1.9×10^{-4}	22.6
H-(cyclohexyl)-CO$_2^\ominus$	β-CD	7.1×10^{-3}	
	Ia-Zn$^{2\oplus}$	5.3×10^{-4}	13.4
	β-CD	1.3×10^{-2}	
	Ib	7.8×10^{-3}	
1,8-ANS	Ib-Zn$^{2\oplus}$	1.9×10^{-3}	6.8
	Ia-Zn$^{2\oplus}$	4.5×10^{-3}	2.9
(naphthalene-COCH$_2$CO$_2^\ominus$)	β-CD	1.4×10^{-3}	
	Ia-Zn$^{2\oplus}$	2.4×10^{-4}	5.8
H-(cyclohexyl)-NH$_2$	β-CD	2.2×10^{-2}	
	Ia-Zn$^{2\oplus}$	4.7×10^{-3}	4.7
NO$_2$-(phenyl)-O$^\ominus$	β-CD	2.1×10^{-3}	
	Ia	1.8×10^{-4}	
	Ia-Zn$^{2\oplus}$	8.3×10^{-4}	2.5
H-(cyclohexyl)-OH	β-CD	2.0×10^{-3}	
	Ia-Zn$^{2\oplus}$	2.4×10^{-3}	0.83

Source: I.Tabushi, N. Shimizu, T. Sugimoto, M. Shiozuka & K. Yamamura, *J. Amer. Soc.*, **99**, 7100 (1977).

a considerably fixed coordination site for the metal binding. Only when the sequential order of several monomer units can be precisely controlled, can this kind of sophisticated function be obtained with polymers. At present, the best method is to combine independently prepared base polymers with low molecular functional sites according to the 'building up' principle.

To answer the question, raised at the beginning of this section, of how to bind phenols or benzoates specifically, we prepared polymers functionalized with the holohost described above. This polymer strongly binds phenols and benzoates, as expected, and extracts these compounds from their concentrated aqueous solution or even from their dilute solution remarkably. After the extraction, only 2–4 ppm of these ions remained unbound in the solution after rather short contact time, as can be seen from Table 3.17.

Table 3.17. *Extraction of traces of hydrophobic anions from dilute aqueous solution by triaminocyclodextrin* $\cdot Zn^{2\oplus}$ *bound to a polymer*

Guest, initial conc. (ppm)	pH	Guest, final conc. (ppm) $Cu^{2\oplus}$	$Ni^{2\oplus}$
$CO_2^{\ominus}$ (naphthalene)	(43)	6.9	2	4
$CO_2^{\ominus}$ (naphthalene)	(43)	6.9	2	4
$O^{\ominus}$ (naphthalene)	(36)	10.0	2	2

Fig. 3.18. Molecular 'recognition' of hydrophobic ketocarboxylates by triaminocyclodextrin $\cdot Zn^{2\oplus}$

$$\Delta G_{T} = \Delta G_{inc} + \Delta G_{cor} \qquad (3.28)$$

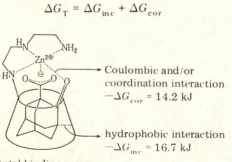

Coulombic and/or coordination interaction
$-\Delta G_{cor} = 14.2$ kJ

hydrophobic interaction
$-\Delta G_{inc} = 16.7$ kJ

total binding energy of double recognition
$-\Delta G = 31.0$ kJ

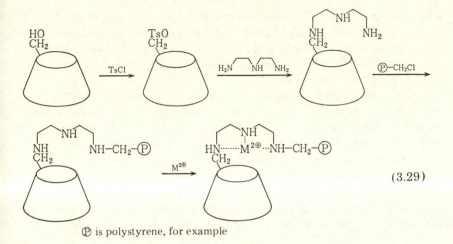

(3.29)

℗ is polystyrene, for example

3.10   Characteristics of enzyme reactions

As discussed above, obtaining even simple enzyme functions requires a great deal of sophistication. At present, this sophistication can only be conveniently achieved by use of a well-designed 'active site' model of rather low molecular weight compounds. Returning to our original question of how to design polymers with enzyme-like activities, we should find out how the native enzymes fulfil their complicated functions by examining some real enzymatic reactions.

3.10.1   *Activation of substrates through specific binding*

Most hydrolytic enzymes possess this characteristic (serine proteases, nucleases, lysozyme and other saccharases). Specific binding of the substrates is carried out both by the surface (e.g. a hydrophobic pocket or a cleft) which 'recognizes' the size and shape of the substrate and also by the points (e.g. hydrogen-bonding site, salt bridge, or metal coordination site) which 'recognize' substrate orientation, stereochemistry or regiochemistry. This profile of the specific binding is common to most native enzymes. Activation of the bound substrates is very often achieved indirectly, e.g. by restriction of the internal rotation of the bound guest or desolvation of the bound guest by the hydrophobic environment of enzymes. Restriction of the internal rotation of the guest facilitates attack by a catalytic group or a bound reagent to the bond of the guest to be converted, and desolvation of the guest leads to enhancement of the reactivity of the guest. Electrostatic stabilization or hydrogen bonding stabilization of the bound transition state is also a commonly adopted mechanism in enzyme catalysis. Direct activation of the bound substrate is important as an alternative mechanism for enzyme catalysis. Serious strain is induced in sugar binding to the lysozyme cleft where one of the glucose residues (D-ring) is forced to take a half-chain form due to repulsion between the sugar and the enzyme wall. The induced strain, serious in a starting material, is greatly relaxed in a transition state where the C–O–C linkage connecting the D-ring with the adjacent ring is appreciably loosened. As a result, the potential increase in the starting material is much larger than that in the transition state, leading to a decrease in the activation energy.

Another typical example of direct activation of the bound guest is metal coordination, which is well known for carboxypeptidase catalysis where coordination of the peptide carbonyl to $Zn^{2\oplus}$ in the active site decreases electron density on the carbonyl carbon and thus accelerates attack of nucleophile on the carbonyl carbon. Electron deficiency of $Zn^{2\oplus}$ at the active site may be understood as resulting from insufficient charge neutralization due to coordination of two neutral ligands and a single anion (Fig. 3.19 and Table 3.18).

Based on these considerations, future synthetic polymers should be functionalized for the specific binding of certain substrates, where the bound

Table 3.18. *Specific binding of a guest by carboxypeptidase*

Substrate	Enzyme	Determining factor for recognition
CO_2^- at C-terminus	$\oplus$ of arginine	Distance between $\oplus$ and $\ominus$ in a hydrophobic environment
Hydrophobic substituent at C-terminus	Hydrophobic pocket	Shape and size of a hydrophobic substituent
Amino acid residue to be cleaved in the substrate	OH of tyrosine	H-Bonding, distance and angle
C=O of a peptide to be cleaved	$Zn^{2\oplus}$	Coordination distance and angle, activation of CO toward nucleophilic attack

Fig. 3.19. Molecular recognition of a peptide substrate by carboxy-peptidase A as a model of sophisticated recognition. (Based on E. Zeffren & P. L. Hall, *The Study of Enzyme Mechanisms*, John Wiley (1973).)

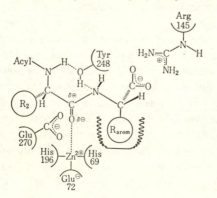

substrates are activated by the polymer. To achieve this, interaction among functional groups on the polymer should be strengthened and specified in order to retain strictly the shape and size of the active sites.

3.10.2 *ES complex formed by specific binding of guests:*
Structural change of an enzyme from its resting form to its
active form by guest binding

Enzyme catalysts may be classified into the following categories, based on chemical grounds:

(*a*) an enzyme activates itself via intraprotein interaction between two (or more) functional groups, as in the so-called 'charge-relay' system of chymotrypsin;

(*b*) a substrate is activated by an enzyme, as discussed above for lysozyme or carboxypeptidase;

(*c*) an enzyme is activated by the substrate binding, e.g. via conformation change, etc.;

(*d*) enzyme catalysis is regulated by a headquarter or a signal compound.

Item (*d*) will be discussed further in a later section. This section will discuss the catalytic action of cytochrome P-450 as an example of category (*c*).

Cytochrome *P*-450 is the name of a large family of enzymes widely distributed in mammals to bacteria. A common characteristic of these enzymes is activation of O_2 molecules with a haem and a reductant (NADH, NADPH). This activation leads to monooxygenation of the substrates via hydroxylation of an alkyl group or epoxidation of an olefinic group. Interestingly, haemoglobin and myoglobin form stable complexes with O_2, and the electron transfer from haem ($Fe^{2\oplus}$) to O_2 is only a side pathway, in marked contrast to O_2 activation by *P*-450, although the local structure of *P*-450 is very close to those of O_2-carrying proteins. Therefore, the marked difference in chemical behaviour between the monooxygenase and the O_2-carrying proteins must be due to the difference in proteins. Even for the O_2-carrying proteins, a considerable charge transfer is usually observed.

$$P \cdot Fe^{II} \ldots O_2 \longleftrightarrow P \cdot Fe^{III} \ldots O_2^-$$

This strongly suggests that the magnitude of the charge transfer is not a determining factor for differentiating between oxygenases and O_2-carrying proteins. In the literature, some discussions are under way about the significance of axial ligation of S^-, but in model experiments we have found that the imidazole ligand is as effective as the S^- ligand for monooxygenation by the $P \cdot Mn^{III}$ (or Fe^{III})–reducing reagent–O_2 system. This finding suggests that the presence of the axial S^- ligand is not a necessary condition for the *P*-450 activity. Therefore, the *P*-450 monooxygenase activity is due to facile protonation of the charge-transfer state by protein and facilitated second electron acceptance. Accommodation of a substrate in the active site should be another requirement.

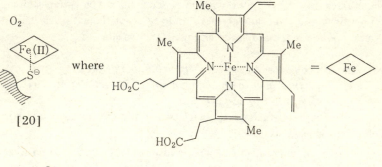

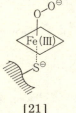

[20]

[21]

$$[21] \overset{H^{\oplus}}{=\!=} Fe(IV)-O-OH \overset{e}{\longrightarrow} Fe(III)-O-OH$$

$$\overset{H^{\oplus}}{=\!=} Fe(III)-O-^{\oplus}OH_2 \longrightarrow Fe(III)-O^{\oplus} + H_2O$$

$$Fe(V)=O$$

[22]   (3.30)

The most probable reaction mechanism of the *P*-450 monooxygenation is shown by equation (3.30). Let us consider the effect of substrate binding on enzyme activation.

(1) Binding of the specific substrate may cause a conformation change of the protein to facilitate electron acceptance from the NADH-reductase system.

(2) When substrate is absent in the active site, the active site wall must be seriously damaged by the potent oxidizing reagent [22] formed there *in situ.*

Therefore, under the activation condition, substrate binding is necessary to protect the enzyme itself.

Based on this information, polymer catalysts having *P*-450 activity can be designed. The difficult problems are: (i) substrate binding at the juxtaposition to the edge of the active species, [22] – if this does not occur, the polymer skeleton will be seriously damaged by [22]; (ii) effective activation of O_2 by smooth electron and proton transfer from the bulk solution – if this does not occur the polymers become simple oxygen carriers.

Assume that we can make a minute hole or 'cavity' in the polymer to which some specific guest molecules are bound. In this bound state, catalytic functional groups keep ideal spatial arrangement for the catalysis. If we have such elegant polymers, the above problems may be promptly solved. Recently, Wulff has developed a new technique to make this kind of artificial active site in artificial polymers.

In Wulff's basic strategy an appropriate bi- (or poly-) functional monomer of rigid skeletal structure (tentatively called a 'signal' comonomer) is copolymerized with common polymerizable monomers (Fig. 3.20). The resulting copolymer as a whole should be rigid by crosslinking. The functional group of the 'signal' comonomer can be cleaved under mild conditions, after which the polymer's rigid skeletal structure is the same as, or at least very similar to, the original skeletal structure. The size and the shape of the 'cavity' thus formed in the polymers by the cleavage are the same as, or very similar to, those of the expected substrate in every respect. Thus, when the copolymerization is over, this rigid copolymer is treated with a reagent capable of cleaving the functional groups of the 'signal' comonomer to obtain the 'specific' cavity. This cavity should have a size and shape exactly the same as, or at least very similar to, those of the specific guest molecules and should have catalytic functional groups at the juxtapositions for certain catalytic reactions. Although the study is still in its early stages, Wulff and his coworkers have mostly concentrated on the specific binding problem, and have already found several interesting points of substrate specificity – specificity toward size and shape of the substrate, including satisfactory stereochemical or chiral 'recognition'. It is hoped that this, and maybe more sophisticated future strategies, will be able to solve the various problems discussed in this section on polymer catalysis.

3.10.3 *Allosteric enzymes*

The regulatory function of certain enzymes mentioned in item (*d*) of the preceding section is one of the most difficult problems to understand from the chemical standpoint. The term allosteric control not only refers to the simplest 'sigmoid' kinetics as seen in the O_2 adsorption by haemoglobin, but also includes the very sophisticated, delicate and complex function of regulating the concentrations of many physiologically important compounds. A certain substrate, product, or any other compound can become a 'signal', which causes

Fig. 3.20. Schematic representation of the basic strategy for preparing an artificial active site in the polymer.

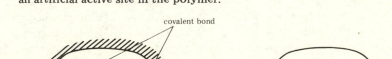

a remarkable or even dramatic change in the enzyme activity due to the conformation change induced by its binding. If necessary, the 'response' to a signal compound appears in a very much amplified way as is discussed in some detail below.

The response to the signal should be precise and delicate. When an excess of starting material is present, the enzyme system should drive the catalytic process forward (or accelerate the forward rate constant) but when an excess of a final product is present (or a shortage of starting material exists), the enzyme system should drive the catalytic process backward (or retard the forward rate constant). The system should also respond to other indirectly related compounds by activating or deactivating the catalytic processes. Now let us consider glutamine synthetase as a typical example of very sophisticated regulatory enzymes.

Glutamine synthetase in the 12-mer of the subunit of 50 000 daltons where six subunits are arranged in a hexagon and two planar hexagons are stacked as a two-storied structure. The most significant feature of this allosteric enzyme is its catalytic action for a wide range of biologically important reactions. Interconversion between glutamate and glutamine is catalysed by this enzyme as is evident from its name. It also catalyses an amino transfer reaction between glutamine and α-ketoglutaric acid.

$$\text{Glu-CO}_2\text{H} + \text{NH}_3 + \text{ATP} \rightleftharpoons \text{Glu-CONH}_2 + \text{ADP} + \text{Pi} \qquad (3.31)$$

$$\text{HO}_2\text{C-(CH}_2)_2\text{COCO}_2\text{H} + \text{NH}_3 \rightleftharpoons \text{H}_2\text{NOC-(CH}_2)_2\text{CHNH}_2\text{CO}_2\text{H} \quad (3.32)$$

Adjusting the equilibrium states of these interconversion reactions controls the concentration of NH_3 in a biological system. Control of the NH_3 level is especially important as this regulates the levels of tryptophan, histidine and many other amino acids. In other words, the above reactions readily lessen any shortage of a certain amino acid by catalysing the corresponding reaction of keto acid with an increased level of NH_3. Similarly, biosynthesis of purine or pyrimidine base is also controlled by the NH_3 level. Moreover, the NH_3 level controls levels of ATP, CTP, GMP, and AMP as well as phosphate, through which levels of sugar phosphates are also controlled.

Interestingly, the catalytic activity of glutamine synthetase depends on the levels of these physiologically important compounds in a very sophisticated and well-designed way as shown in Fig. 3.21. The activity also depends on the concentrations of physiologically important metal ions such as $\text{Mg}^{2\oplus}$ and $\text{Mn}^{2\oplus}$. This activity control becomes more complex with the intervention of the adenylation-deadenylation equilibrium of the enzyme. The question arises of how the enzyme can change its activity so sensitively toward the concentration changes of almost all physiologically important compounds, and the answer is by allosterism. An enzyme has many different recognition sites $B_1, B_2, \ldots, B_n$ for many different physiologically important compounds, $S_1, S_2, \ldots, S_n$, respectively, where the binding of a certain bioactive compound, S_i, to the binding site, B_i, causes a remarkable structural (mostly conformational) change in the subunit to which B_i belongs. Since the intersubunit interaction (may be ionic, hydrogen-bonding and/or hydrophobic) is very strong, this structural change causes a corresponding structural change in the rest of the subunits. Thus, the catalytic constant of the reaction between bioactive compound S_j and the corresponding binding site B_j changes due to the structural change of the subsite to which B_j belongs. This is caused by the binding of the independent substrate S_i to the independent site B_i. The activity change is not limited to acceleration (shown by $\oplus$ in Fig. 3.21), but deceleration (shown by $\ominus$) often being caused.

At present, several recognition sites are known to which several bioactive compounds bind competitively:

> For B_1, glutamate; competition with glycine, alanine, tryptophan, CTP
> For B_2, $NH_4^{\oplus}$; competition with histidine, glucosamine-1-phosphoric acid
> For B_3, AMP and carbamyl-1-phosphoric acid

A careful look at the entire scheme of Fig. 3.21 shows that a slight change in the concentration of the bioactive compound causes a dramatic change in the concentration of the final products, glutamine and α-ketoglutaric acid, which play key roles in controlling the NH_3 level. As this amplification reminds us of either an avalanche caused by a tiny stone or a waterfall originating from water droplets in a mountain, it is called the 'cascade control' mechanism.

Fig. 3.21. Cascade control mechanism by glutamine synthetase of *E. coli*. (From P. D. Boyer, *The Enzymes*, 3rd edn, vol. 10, p. 791, Academic Press (1974).)

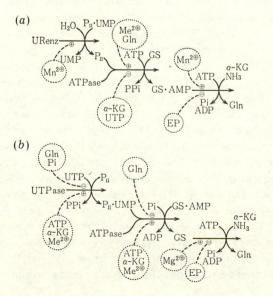

Assume a slight increase in the α-ketoglutaric (KG) acid concentration. The simplest effect of this increase is a direct concentration effect on the KG $\rightleftharpoons$ Gln equilibrium, where the 'response' of the Gln increase to the KG increase is 1:1. However, KG increases the Pu · UMP concentration (response 1: many) which catalyses the formation of GS [1: (many)^2], and GS catalyses the conversion KG → Gln. Therefore, the concentration of Gln increases by a factor of (many)^3 per single KG increase. A careful look at the 'negative' catalytic effect also shows that the present allosterism is very elegant and smooth.

This type of allosterism is a very important one in a variety of regulation mechanisms in living systems. Thus, one of the versatile future aspects in using polymer catalysts for practical purposes is to synthesize excellent regulatory polymers. They should be useful not only in chemical industries but also as a molecular element for information storage, information transfer or information judgement.

3.11   Examples of polymer catalysts having enzymatic activities

This last section briefly surveys literature on the preparation of polymer catalysts with enzymatic activities. Polyvinylimidazole prepared by Overberger and his coworkers is very important as a successful example of preparation of polymer catalysts having hydrolytic enzyme activity. The basic strategy for the preparation of relatively simple polymer catalysts with enzyme activities seems to be still based on this Overberger's concept of statistical formation of an 'active site' by binding two (or more) functional groups at a certain distance and a certain angle. However, remarkable increases in rate acceleration or binding (k_{cat} or $1/K_m$) have been observed in later studies. Although some of the presently known polymer catalysts show rate constants comparable to or even larger than those of certain native enzymes, most of the mechanisms of various native enzymes have not yet been successfully 'mimicked'. For example, organic esters, especially hydrophobic esters, are very readily hydrolysed by polyvinylimidazole or many other functional polymers, but the hydrolysis proceeds via nucleophilic catalysis by an imidazole group. In contrast, serine proteases such as chymotrypsin utilize an imidazole group of His-57 as a general base catalyst, which donates an electron pair to the oxygen atom of Ser-195, an active nucleophile. As a result, the imidazole group remains unchanged during the catalytic process and can be utilized again in the successive deacylation reaction where the imidazole acts as an effective general base and a water molecule is activated by donation of an electron pair from the imidazole. In this way, deacylation of chymotrypsin is also accelerated remarkably and a single imidazole group can be used repeatedly (called efficient 'turnover'). However, most simple polymer catalysts do not show efficient 'turnover' due to a slow deacylation step. The so-called bell-shaped pH dependence, another characteristic of native enzymes, is not reproduced by a simple polymer catalyst such as polyvinylimidazole, where the catalytic constant monotonously increases with the pH increase to reach a maximum value. Unfortunately, the substrate-shape specificity has not yet been success-fully 'mimicked' by simple polymer catalysts, although it is one of the most significant characteristics of native enzymes. The introduction of hydrophobic (lipophilic) groups into polymers as pendant groups increases the affinity toward hydrophobic guest molecules in general. However, even in this case, strict discrimination between a certain specific guest molecule and nonspecific guest molecules of similar structure is practically impossible.

In spite of these disadvantages, polymer catalysts have a remarkable advantage in that once prepared appropriately they can be easily applied to practical purposes. Another advantage is their simple preparation based on the idea of: (functional groups in the active site) + (polymer as a supporting material and as a hydro-phobic medium). Several simple polymer catalysts were reported as displaying the bell-shaped pH profile and enormous acceleration. Very often the accelera-tion was of an order of magnitude similar to the native enzyme (for nonspecific substrates). Thus, the activity of some artificial polymer catalysts seems to closely approximate that of native enzymes.

A series of hydroxamic acid–imidazole–ammonium polymers were prepared by Kunitake and coworkers. In these polymers, the nucleophilic catalysis of hydroxamate is assisted by imidazole (or imidazole acts as a general base catalyst). In other words, appropriately functionalized polymers hydrolyse substrates via a mechanism very similar to that of the native enzyme. Acyl hydroxamate, an intermediate corresponding to the acyl enzyme, is readily hydrolysed by imidazole catalysis, resulting in reasonably efficient 'turnover'. This characteristic is also exactly what was observed for native enzymes.

These reported polymer catalysts are excellent in many respects and allow the optimism to imagine that more sophisticated enzyme activities will be mimicked satisfactorily in the not too distant future by polymer catalysts. However, even for these successful examples, note that much simpler systems constructed with low molecular weight compounds also exhibit similar activities. For example, Tabushi showed that a catalytic amount of hydroxamate having an appropriate hydrophobic tail can be used with cetyltrimethylammonium bromide micelles as a potent catalyst having enzyme activity by use of hydroxamic acid, which itself was found to be a good enzyme model from the mechanistic point of view. The catalytic constant of the hydroxamate polymer prepared as an extension of these catalyst systems of low molecular weight compounds, is of the same order of magnitude as the micelle system discussed above.

Many polymer chemists are dreaming of the existence of the polymer effect in enzyme-like catalysis which is only operative for high molecular weight catalysts. But its existence has not yet been demonstrated, and we will have to wait to see if the dream comes true.

Many other polymer catalysts are prepared as enzyme models; polymer catalysts for oxidation–reduction, polyions, phase-transfer polymers, polymers having nucleic bases as pendant groups, polymers which convert solar energy

to chemical energy ('artificial plant') and polymer–metal complexes. Most are
discussed in other chapters and will only be mentioned briefly here. Let us
examine a few recent examples which show some unique and significant
characteristics.

   A phase-transfer reagent is a kind of surfactant which has a hydrophilic site
as well as a hydrophobic site. A hydrophobic site usually consists of two or more
medium-to-long alkyl chains, and has a steric requirement to destabilize the
micellar state (a surfactant having a single alkyl chain stabilizes the micellar state).
A hydrophilic site is usually an ion or a set of neutral coordinating groups such
as crown ether or cryptand. These phase transfer reagents are easily immobilized
on the polymer, giving phase transfer polymers. Here are some typical examples.

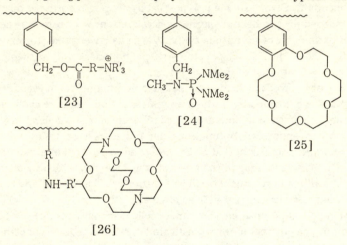

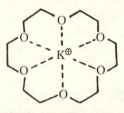

In polymer-bound crown ether, the macrocyclic hexaether coordination site,
called a crown ether, has a strong affinity for $K^\oplus$ and readily cleaves the ionic
bond of $K^\oplus X^\ominus$, even in a hydrophobic environment, by producing the strong
coordination complex [27]. Since crown ether is a neutral ligand, complex [27]

is still positively charged and requires the presence of a counter anion, $X^\ominus$, in its
vicinity. Thus, polymer bound crown [25], dissolves KX in a hydrophobic
medium in spite of the fact that KX otherwise has a negligibly small solubility
in the medium. Note the important fact that anion $X^\ominus$ is poorly solvated compared with that in an aqueous or a polar medium. Also, the ionic interaction
between $X^\ominus$ and crown $\cdot K^\oplus$ is much weaker than the direct interaction,

$X^{\ominus} \ldots K^{\oplus}$, in a less polar medium. Therefore, $X^{\ominus}$ is considerably destabilized or, in other words, remarkably activated. If polymer-bound crown [25] is used in a hydrophobic medium, a certain S_N reaction of the type:

$$KX + RY \longrightarrow RX + KY$$

should be greatly accelerated via exactly the same mechanism observed with monomeric crown ether. However, using polymer crown ether has a great advantage over monomer crown ether: the desired product RX (in a hydrophobic medium) can be spontaneously and instantaneously separated from another product KY (bound to a solid polymer) without any difficulty when polymer crown ether is used. Therefore, many polymer-bound crown ethers and related compounds have been prepared, mostly by Regen & Montanari. At present, crown ether or a related grouping is usually introduced to the polymer as a pendant group. Reactions catalysed by the polymer-bound crown ethers are spread over a wide range of organic reactions, including S_N2-type reactions of $Hal^{\ominus}$, $HO^{\ominus}$, $RO^{\ominus}$ and carbanion, or dichlorocarbene reactions, decarboxylation

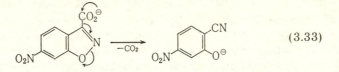 (3.33)

reaction (3.33). Asymmetric induction or asymmetric selection have also been investigated using polymer catalysts.

With all of these important polymer catalysts, the experimental procedures are very simple compared with common solution reactions. The remaining problem is that the formation of the hydrophobic environment around the active site of the phase transfer reactions cannot yet be satisfactorily done. This may be solved by developing a triphase system (aqueous, polymer catalyst, organic) as proposed by Regen.

4

POLYMERS HAVING ENERGY-CONVERTING CAPABILITY

SHIGEO TAZUKE

When one looks at energy-resource problems, there is no doubt that they reduce to problems of technology and engineering to be developed on the basis of chemistry and physics. Any technology relevant to transportation, storage and transformation of energy can be placed in the category of energy conversion. This is very fertile ground in which chemists, physicists and biologists can contribute greatly regarding contemporary concern about energy resources.

While the significance of science lies in the understanding and elucidation of the principles and causality of phenomena, engineering aims at establishing systems by combination of various phenomena and principles that are meaningful and influential in our actual society. Consequently, the difference between engineering and science lies in the procedure used for tackling a problem and, more fundamentally, in the standards used for evaluating research output, rather than in the research subject itself. Let us confirm that the energy-conversion problem is an engineering-oriented problem and consider the difficulties of polymers having an energy-converting capability.

If we overlap conversion as an integrated technology with polymer science as a branch of applied science, the overall effect is going to be a practical and utilitarian one. However, in practice, no one knows what field is covered by the name 'energy-converting polymers'. As a matter of fact research in this area has not yet reached a stage where practical evaluation may be possible. Nevertheless, there are several polymer materials that show potential. The aim of this chapter is to assemble the fruits of basic research and to think over future possibilities.

4.1 Various energy-converting processes and their relevance to polymer materials

To begin with, an overview of energy conversion in general is presented. Any processes relevant to conversion of energy between different forms are defined as 'energy conversion' in a narrower sense. However, besides conversion of energy between different forms, transformation of energy density (transformation of voltage, wavelength conversion of light, etc.), spatial transformation (transportation of energy, power transmission, communication, etc.), and transformation along a time scale (energy storage, battery, flywheel, etc.) may be included in the category of energy conversion in a wide sense. These subjects,

not relevant to interconversion of energy 'forms', are excluded from the present
discussion.

Various energy conversion processes, excluding nuclear energy, are shown in
Table 4.1. As shown below energy conversion between any two forms of energy
is *possible*, providing one neglects the conversion efficiency. From the viewpoint
of energy resources, however, energy conversion processes of significance are
confined to those which enable the conversion of lower grade energy to higher
grade energy with a high efficiency. This does not mean that other processes
are of no significance. From the standpoint of sensors for detecting information,
many processes shown in Table 4.2 are practically important, and furthermore
polymer materials take on a major role in certain cases.

In contrast to uses as sensors, polymer materials are still in the position of
supporting materials for orthodox large-scale energy conversion. Since the purpose
of large-scale energy conversion is to provide the technology to convert low grade
energies, which are amply available but difficult to store and transport, to high-
grade energies in more usable forms, restrictions are more than those in the case
of sensor application. Thermal energy is easily obtainable by photoabsorption,
combustion or the Joule effect, and hence it is of lower grade than photo-,
chemical or electromagnetic energy. Consequently, the reverse conversion
processes are important. Photoenergy is not a suitable form to be stored or
transported; it is more easily available and of lower grade than chemical or
electromagnetic energy. Mechanical energy is available in Nature in the forms
of wind and hydrodynamic powers. As a consequence, a dynamo to convert
mechanical to electromagnetic energy is a key device. Polymers are important
here but only as supporting materials, such as insulators, and constituents of
lubricants. If superconducting polymers could be developed, one would expect
a spectacular improvement in energy conservation efficiency.

In an MHD (magnetohydrodynamic) generator, aiming at direct conversion
of heat to electricity and also in the combination of Peltier effect–Seebeck
effect to transfer heat via heat $\rightarrow$ electricity $\rightarrow$ heat conversion, polymer
materials have not been much used. In the conversion of electricity to chemical
energy, polymers may be useful as separators in batteries and electrolysis vessels
but not as active materials. Electricity $\leftrightarrows$ chemical energy conversion is not significant as regards energy-resource problems.

A very recent advance to use conductive polymers as electrodes of primary
and secondary batteries is worth noting. When polyacetylene composes both
anode and cathode in the presence of an electrolyte such as $Bu_4^n N^\oplus ClO_4^\ominus$ or
$Bu_4^n N^\oplus PF_6^\ominus$, the electrodes are oxidized and reduced electrochemically. The
chemical energy stored in the different oxidation states of polyacetylene can be
released as electricity. This means that a completely organic rechargable battery
is now possible (D. McInnes, Jr., M. A. Drug, P. J. Nigrey, D. P. Nairns,
A. G. MacDiarmid & A. J. Heeger, *Chem. Commun.*, 317 (1981)). The capacity
of the battery per unit weight is very large.

Table 4.1. *Conversion among different energy forms*

From \ To	Mechanical	Heat	Electromagnetic	Photo	Chemical
Mechanical	Simple machine, like a lever	Rubbing	Dynamo Piezoelectric effect	Triboluminescence	Reverse mechanochemical
Heat	External combustion engine	Heat pump	Seebeck effect Thermal electron ejection	Temperature radiation	Endothermic reaction
Electromagnetic	Motor	Peltier effect Joule effect	Transformer	Electroluminescence discharge Luminescence	Electrolysis Electrode reaction
Photo	Light pressure Photomechano-chemical	Photoabsorption	Photovoltaic effect Photoelectric effect	Laser Fluorescence Phosphorescence	Photosynthesis Endothermic photoreaction
Chemical	Mechanochemical Internal combustion engine	Exothermic reaction	Primary battery Secondary battery Fuel cell	Chemiluminescence	Chemical reactions in general

Using photoenergy as the primary energy resource, the conversion processes to electricity and heat have been well developed. As active materials for photo-cells, inorganic semiconductors, represented by silicon, are thought to be most promising at the moment. Nevertheless, there seems to be a good possibility that semiconducting polymers will participate in future. Conversion of photo-energy to thermal energy is easy in principle and the problems are reduced to engineering and economic aspects. Here polymers are once again used as supporting materials. Polymers are more closely involved in photoenergy → mechanical energy conversion and photoenergy → chemical energy conversion as will be discussed in §4.4.

Characteristics of chemical energy are the ease of storage and transportation. The best example utilizing these characteristics is the chemical → mechanical energy conversion by muscles, which are typical energy-converting polymers. Apart from irreversible combustion, combination of endothermic photochemical reactions, such as isomerization and dimerization, with thermal back reactions is an interesting subject relevant to the storage of solar energy as chemical energy. Polymer materials are expected to be used as polymer sensitizers and polymer catalysts. Chemical → electrical energy conversion by means of a primary battery and a fuel battery is also an immediate problem. Here again, polymer materials play only minor roles. Chemiluminescence is an example of direct chemical – photoenergy conversion. A certain thermal decomposition or redox reaction of high energy compounds such as dioxetanes proceeds via electronically excited states (usually singlet excited states) which are luminescent. These phenomena find use as fishing and emergency lights. This is a high grade → low grade energy conversion and of minor significance.

Table 4.2. *Roles of polymer materials in energy conversion devices as sensors*

Energy conversion	Examples of polymer materials
Mechanical → electromagnetic	Piezoelectricity, sound pressure → electricity conversion in microphones, and pick-ups (poly(vinylidene fluoride) etc.)
Thermal → electromagnetic	Switching device by means of plastic thermistor
Photo → electromagnetic	Imaging system by photoconduction (poly(9-vinylcarbazole) etc.)
Ionizing radiation → photo	Detection of ionizing radiation by plastic scintillator
Photo → chemical	Imaging system by photopolymers
Chemical → mechanical	Hair-hygrometer
Chemical → electromagnetic	Various sensors with functionalized electrode or enzymatic electrode

The present energy problem originated from the shortage of fossil fuel, namely the shortage of chemical energy. Although a variety of unused chemical energy resources are conceivable, all of them stem from solar energy (photo- and thermal energy). Wind and hydrodynamic powers are from solar energy as well, as shown in Fig. 4.1. When we realize the facts that all available energies in the earth are of solar origin – except for a few instances of atomic energy, geothermal energy, the westerly wind due to the rotation of earth and so on – it is a serious problem how to fit our energy conversion system into the solar energy flow chart of Fig. 4.1. In particular, the photosynthetic system driving photo- → chemical energy conversion and the muscle system for chemical → mechanical energy conversion are primarily made possible by most refined functions of polymers (or, more generally, molecular aggregate systems). Let us borrow knowledge from Nature and survey possible polymer systems which could contribute to solving energy-resource problems.

4.2   Information from biological energy conversion

Among the various chemical processes shown in Fig. 4.1, photosynthesis is the most subtle system. Also, the movements of animals using photosynthesis products as fuel (namely, direct conversion of chemical → mechanical energy) are instructive. In addition, the delicate mechanisms of the senses represented by vision, audition, smell, taste and touch are intriguing. The primary stimulations

Fig. 4.1. Material and energy circulation in Nature.

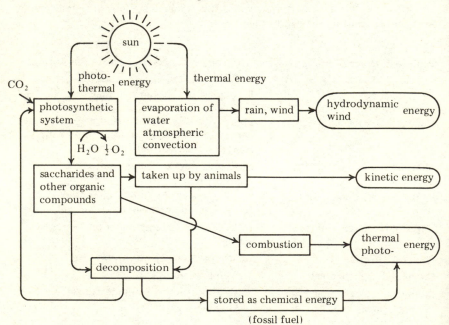

(fossil fuel)

are given in the form of photo- (vision), mechanical, i.e. sound pressure (audition), chemical (taste and smell), mechanical and thermal (touch) energies. These stimulations are eventually converted to electrical signals to excite nerves. Although studies to understand the problems of the senses on molecular level have been undertaken, our present understanding is far from satisfactory.

4.2.1 *Mechanism of muscle movements*

Before modern molecular theories of muscle contraction had been proposed, several models to explain the correlation of muscle strength with muscle length and loading rate were presented. The mechanical model (Fig. 4.2*a*) is an attempt to express measured muscle contraction based on the constants for various elements. The model is the combination of contraction element (x_c) and elastic element (x_e) which are evaluated from the muscle strain–stress chart, the time-dependent iso-strain contraction behaviour, and others. The distorted chain model (Fig. 4.2*b*) is based on the assumption that muscle distortion is brought about by the chemical bondings in small contractile elements which constitute

Fig. 4.2. Various muscle models. (*a*) Dynamic model. A: normal state, B: contraction with a constant stress, C: contraction with a constant length, *p*: load, x_c: length of contracting element, x_e: length of elastic element, *s*: contraction. (*b*) Distorted chain model. α: contracted chain, β: related chain. (*c*) Sliding model. A and M represent the active parts of actin and myosin, respectively. The arrows show the direction of movement. (From *Molecular Theory of Biological Functions*, Seibutsu-butsurigaku Koza, vol. 6, Yoshioka-Shoten (1965), Figs. 25–27.)

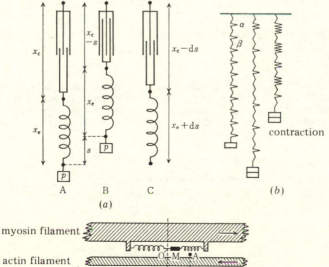

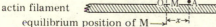

(*c*)

a muscular tissue unit. Each element is a distorted chain. Taking no intermediate states into account, the probability of contracted (α) and extended (β) states as a function of load, and the ratio of lengths ($\alpha:\beta$), are assumed to explain muscle movements. The physicochemical model assumes that contracted and extended states are interrelated via chemical reactions, and the concept of activation energy is introduced.

These models provide, as it were, an explanation for the sake of explanation, regardless of the actual movements of muscles. Contrary to these earlier models, direct observation of the fine structure of muscular tissue and its dynamic structure has unequivocally shown that myosin (A fibre) does not display any morphological change during muscle contraction. Based on this observation a new theory was presented which claims sliding of actin filaments into the spaces between myosin filaments. The early model by H. E. Huxley (Fig. 4.2c) cannot explain muscle contraction well. However, the intuition to assume binding between actin and myosin which pulls each filament together has been recognized as correct by subsequent researchers.

The muscle-constructing molecules have now been well characterized. Actin consists of protein of mol. wt $=6 \sim 7 \times 10^4$ as the minimum unit and exists in various lengths, whereas myosin has a mol. wt of $5 \sim 6 \times 10^5$ and takes a helical form as shown in Fig. 4.3. The head and the tail parts are said to be fixed on actin filaments, and a swinging motion causes muscle contraction. Along the helical part, there is a weak segment that is susceptible to enzymatic scission. The helix structure would be loosened at this weak position which controls the swinging motion. Muscle contraction is triggered by a membrane potential change at the muscular sheath and begins with the ejection of $Ca^{2\oplus}$ into the muscular tissue. The driving energy is supplied by the reaction: $ATP \rightarrow ADP + H_3PO_4$. There is a fast part in muscle contraction behaviour corresponding to the spring

Fig. 4.3. Structure of muscle filaments. (*a*) Myosin (LMM: L-meromyosin, HMM: H-meromyosin), (*b*) F-actin (double helix of spherical G-actin). In the case of muscle contraction, actin and myosin filaments, which are mutually layered, slide into each other triggered by a concentration change of $Ca^{2\oplus}$ ion.

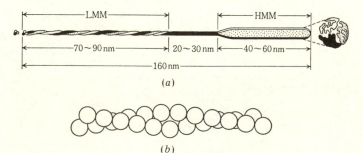

(*a*)

(*b*)

segment in the mechanical model. The elasticity of myosin helix would be responsible for this. The following slow contraction is supposedly coupled with energy
release from ATP. The coupling mechanism is not yet clarified.

The amount of muscle contraction is usually $\sim 20\%$ and at maximum 40% of
its normal length. The maximum tension is $6 \sim 10\,\mathrm{kg\,cm^{-2}}$ and $1.8 \sim 2.2\,\mathrm{kg\,cm^{-2}}$
for human muscles and frog's muscles, respectively. Muscular models with
synthetic polymers (§4.3) are based on the classic models requiring contraction
of molecular chains. It is of interest why Nature has chosen the sliding mechanism rather than the polymer-chain contraction mechanism. Possible reasons
are (i) the fractions participating in active movements are less for the sliding
model than those for the chain contraction model and therefore muscle contraction will be facilitated and (ii) the chain contraction mechanism will not be
adequate when a thick and strong bunch of muscular fibres is to be contracted
with a short response time. For the chain contraction system to be activated,
reaction of all molecules is necessary and response is slow. In fact, all models
described in §4.3 respond slowly, and many tricks are to be still learned from
Nature.

4.2.2   Mechanism of photosynthesis

The structure of a chloroplast in plants is shown in Fig. 4.4. The small
units are called grana which consist of stacks of layered membranes (lamella
system). Each layer takes the form of a closed envelope called a thylakoid. The
thickness of the envelope membrane is about 3 nm. The remaining space inside
the chloroplast is called the stroma in which the CO_2-reducing cycle is located.
All the systems are enclosed in a membrane to construct a chloroplast of a few
micrometres in diameter.

A schematic representation of photosynthesis in the lamella system is given
in Fig. 4.5. This is an oversimplified picture, and in fact the flow of electrons

Fig. 4.4. Structure of chloroplast. G: grana, S: stroma, L: lamella,
SL: stroma lamella, T: thylakoid membrane, CM: chloroplast membrane.

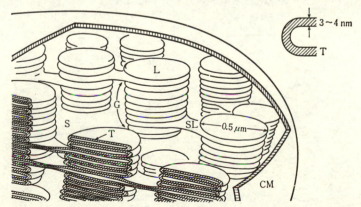

between H_2O and P680, System II and System I, P700 and $NADP^\oplus$ is carried out by many electron transporting cycles. The aim of showing this well-known picture is to establish the purposes of photosynthesis. These are (i) coupling of oxidation of water to reduction of NADP, (ii) photophosphorylation to produce ATP coupled with electron transfer in Systems I and II, and (iii) fixation of CO_2 with the reducing power of NADPH and the energy stored in ATP. If we could conduct photofixation of CO_2, we would not care about the intermediate chemical processes. Aiming at an artificial energy conversion system by modelling and mimicking photosynthesis, a carbon copy of what Nature does, is not possible and also unnecessary. What is required is the technological and practical attitude to utilize the concept of photosynthesis; it is not the object to elucidate completely the mechanisms of action of chlorophyll, or to approach the total understanding of photosynthesis.

This attitude may be compared with the following story. Let us consider a work-horse as a subject for research. One person conscientiously measures various characteristics of the horse such as weight, height, muscle strength, output power and so forth and tries to elucidate why the work-horse is so strong. Another begins by examining the utility of the work-horse. If it is for transportation, he will then estimate the horse power equivalent to the work-horse and try to replace it with a more convenient cargo truck or something. The former is a mechanism-oriented approach whereas the latter is function-oriented. According to the latter approach, our human beings have been replacing the natural energy of man or animal power, wind and hydrodynamic power with artificial mechanical power machines since the eighteenth century.

How useful is photosynthesis as a model of an energy-converting device? The photosynthetic system involves the following subtle tricks.

(1) Many electron-transport cycles are driven by photoexcitation. There are two electron-transporting pumps named Systems I and II. These pumps

Fig. 4.5. Schematic expression of photosynthetic reactions. PQ: plastoquinone, Cyt f: cytochrome f, PC: plastocyanin, Fd: ferredoxin.

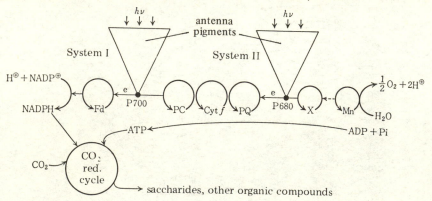

possess absorption bands matched to the solar spectrum. The principle of
photoinduced electron transport is, simply speaking, to increase both reducing
and oxidizing capabilities by electronic transition from the highest occupied
level to the lowest vacant level. The resultant half-occupied level having a higher
energy is a good electron donor whereas the lower-lying half-occupied level is
strongly electron-accepting.

(2) There is a photon-harvesting effect of antenna pigments (Fig. 4.6a).
According to the concept of a photosynthetic unit, the molecular aggregate

Fig. 4.6. Role of antenna pigments. (a) energy transfer in chlorophyll
(this does not mean that the pigments in the thylakoid membrane are
arranged regularly). (From K. Shibata *Kagaku Sosetsu*, **12**, 57 (1976)
ch. 4). (b) Energy migration in synthetic polymers (non-directional).
(c) Energy migration in micelles (non-directional). S: sensitizer segment.

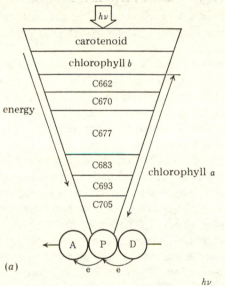

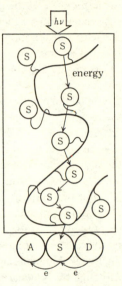

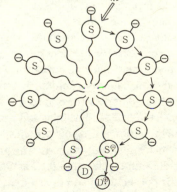

contains one reaction centre of chlorophyll per 200 antenna pigment molecules. The pigments in Systems I and II are not of a single species, and consist of a number of dyes having slightly different absorption spectra. Consequently, energy transfer proceeds from the antenna pigment having the shortest wavelength absorption to those having longer wavelength absorptions and eventually to the reaction centre of chlorophyll. This process is a singlet energy transfer. Assuming the distance between chlorophyll molecules to be 1.8 nm, about 300 steps of energy transfer are estimated to proceed within the excited state lifetime. Even if the ratio of active centre to antenna pigment is 1: 200, energy transfer to the reaction centre is an efficient process. The problem is the ordering of the pigment molecules. Chlorophyll molecules are bound to protein *in vitro* and exist as dye–protein complexes. When the dye molecules are made free by surfactant treatment, the absorption spectra exhibit wavelength shifts, indicating the importance of the polymer matrix. There are as yet many unsolved problems such as possibility of energy transfer between Systems I and II, independence of each photosynthesis unit, sharing of antenna chlorophyll among multiple reaction centres, and so on.

(3) Energy is stored by ADP $\rightarrow$ ATP conversion. The driving force of photo-phosphorylation is the difference in proton concentrations inside and outside the thylakoid membrane, brought about by photoexcitation. Expressed as ΔpH, the required minimum value is 3.5 whereas the photochemically induced pH difference is 3 and slightly deficient. To compensate this shortage, a photoinduced electric field gradient of $\sim$100 mV is estimated. This assumption is supported by the similarity of spectral changes of chlorophyll under photoirradiation and that under an electric field. The conclusion has not been obtained whether the coupling of phosphorylation to the electron-transport system through the thylakoid membrane is mediated by high-energy chemical bondings or by high-energy physical states. As depicted in Fig. 4.7, large particles (17.5 nm) exist on the outer surface and small particles (11 nm) attach to the inner surface of the thylakoid membrane. They are thought to be Systems I and II, respectively. Although the schematic expression of the thylakoid membrane as the key part of photosynthesis may be given, mutual arrangements of the constituents to couple redox cycles are not yet deciphered.

(4) Oxidation and reduction sites are separate. When electron transfer between electron donor and electron acceptor provides an ion radical pair, the reaction efficiency is generally poor unless back electron transfer is suppressed in order to maintain charge separation. The thylakoid membrane is a charge-separating membrane as well as a physical barrier for oxidized and reduced species. The composition of the lamella indicates that lipids and proteins are the main constituents. Besides a small fraction of enzyme proteins, the major part consists of structural proteins which are bound to lipids. The proteins contain hydrophobic amino acids, acidic amino acids, and their amide forms as major components whereas cysteine is not included and the content of basic amino acids is

low. These analytical data suggest that the membrane is built by hydrophobic interactions, and hydrogen bonding contributes in the interior of the membrane. The content of lipid in the lamella is nearly comparable to that of protein. The lipids involved are fatty acids, sterols, waxes, phospholipids and sulpholipids; 94 ~ 96% of the total fatty acids is the highly unsaturated linolenic acid. Some of the lipids possess $-OPO_3^\ominus$ and $-SO_3^\ominus$ acidic groups. The membrane is consequently expected to be organized by hydrophobic and hydrogen bonding interactions and to be soft. Lipids containing linolenic acid are highly hydrophobic and of low melting point. The acidic groups keep the inside of the membrane weakly acidic.

4.3   Conversions to mechanical energy by polymers

4.3.1   *Chemical energy → Mechanical energy*

Polyelectrolytes change solubility and chain conformation depending upon pH, salt concentration and dielectric constant of the medium, which reflects on the size of solid polymer. By utilizing this principle, the contraction or expansion of solid polymer as functions of salt concentration and pH makes chemical energy → mechanical energy conversion possible. Many years ago, A. Katchalsky & W. Kuhn reported the possibility. (W. Kuhn, A. Katchalsky & H. Eisenberg, *Nature, Lond.* **165**, 514 (1950).) Since the muscle contraction mechanism had not been clarified at that time, their experiments gained a good reputation as a muscle model. Although the difference of their model from real muscle motion became obvious later on, it is intuitive that they combined the density change of polymer due to the differences in environments with the idea of energy conversion. The relative viscosity of a dilute aqueous solution of

Fig. 4.7. Electron transport across a thylakoid membrane and an $H^\oplus$ concentration gradient outside and inside the membrane. (From K. Shibata, *Kagaku Sosetsu*, **12**, 68 (1976), ch. 4, Fig. 20.)

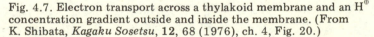
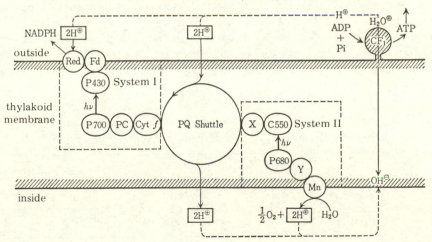

poly(methacrylic acid) increases by a factor of 2000 if half of the carboxylic acid is neutralized. Poly(acrylic acid) crosslinked by ester formation with glycerol or poly(vinyl alcohol) (PVA) could change its size by 300%, depending upon the degree of neutralization. Even with the load 5000 times more than the weight of polymer, about 30% expansion–contraction can be reversibly conducted. Using a sample of 0.1 mm thickness, the response time is less than 1 min. Nevertheless, a very fast response, comparable to that of muscle, is not expected since pH change in the film is a diffusion-controlled process. There are a number of examples of contraction–expansion accompanied by pH changes such as in copolymers of methacrylic acid with divinylbenzene (1~4%) and phosphorylated PVA fibre.

A comparison of polyelectrolyte fibre with muscle is shown in Table 4.3. The PVA–PAA system cannot exhibit oriented contraction and is therefore laminated with a PVA layer that does not change its size with pH change. By means of this sandwich structure, the lateral deformation is suppressed by PVA, and longitudinal deformation alone is possible. From the viewpoint of function, the model is a step towards simulating muscle.

In short, this principle is based on the fact that polymer chains assume a suitable equilibrium conformation to balance the free energy change due to the change in the degree of neutralization and ionic dissociation. The reverse process is also possible: when the polymer chain conformation is altered by an external force, the free energy of the system is affected, which reflects on the pH of the solution. This is a reverse mechanochemical process(mechanical energy → chemical energy). For the PVA–PAA system, the change in pH is only 0.3 at best.

Using the same PAA, the relation between the degree of neutralization and the degree of chain deformation differs for the Na salt and for polyvalent metal salts. The salts of Cu^{2+}, Ca^{2+}, Mg^{2+}, Ag^+ and others insolubilize PAA and consequently, the volume contraction effect is prominent. Alternative addition of Ca^{2+} as a crosslinking agent and EDTA–Na salt as a masking agent for Ca^{2+} can contract and expand the PVA–PAA fibre successively. This is of interest in connection with the effect of Ca^{2+} triggering muscle contraction in biological systems. These artificial polymer contractions are due to intra- and intermolecular crosslinking by ionic or hydrogen bonding. The contraction – expansion cycle can

Table 4.3. *Comparison between muscle and fibre consisting of 80% poly(vinyl alcohol) (PVA)–20% poly(acrylic acid) (PAA)*

	Muscle	Artificial fibre
Tensile strength (kg cm^{-2})	4 ~ 12	4 ~ 12
Contraction (kg cm^{-2})	4 ~ 12	3 ~ 5
Output (J g^{-1})	$60 \sim 80 \times 10^{-2}$	30×10^{-2}

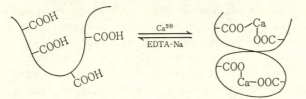

also be achieved by coordination bond formation, a demonstrated by Kuhn for the PVA-$Cu^{2\oplus}$ system. Since then, various combinations of PVA derivatives, polyvinylamines and others, with a number of metal ions have been examined.

Redox reactions can also be used. A copolymer of vinyl acetate with *N*-allyl-barbituric acid is converted to a copolymer of vinyl alcohol (80%) with allyl alloxane (20%). The alloxane copolymer carries out contraction–expansion ($\sim 20\%$) during the redox process of alloxane $\rightleftharpoons$ dialuric acid. The previously mentioned PVA-$Cu^{2\oplus}$ polymer can be expanded by destroying the chelate by $Cu(II) \rightarrow Cu(I)$ reduction. Since $Cu(I) \rightarrow Cu(II)$ oxidation proceeds spontaneously in air, a reversible expansion-contraction cycle could be designed.

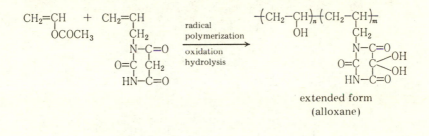

extended form
(alloxane)

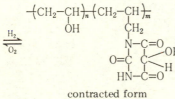

contracted form
(dialuric acid)

Besides Coulombic repulsion, solubility change and crosslinking by salt or chelate formation, changes in crystallinity and higher structures can be used in energy conversion devices. Protein crystals sometimes dissolve by addition of salts. When collagen is dipped in LiBr solution ($5 \sim 6$M), about 8% contraction in volume results. It is observed that the crystal melting point drops suddenly at the salt concentration region responsible for the volume change; this is an indication that the contraction-expansion is brought about by the crystal melting–recrystallization process. A mechanochemical engine stemming from this principle was proposed, as shown in Fig. 4.8. The force generated by belt contraction in the contraction bath is conveyed to C which has a large diameter and is driven anticlockwise. When the engine is actually driven by the combination of collagen crosslinked by formaldehyde with two LiBr bathes ([LiBr] = 11.25 M and 0.3 M for contraction and expansion, respectively), B rotates at a rate of 40 rpm with an output energy comparable with frog muscles. The concentration difference of the two salt baths decreases gradually and the engine eventually stops.

4.3.2 *Thermal energy → mechanical energy*

The various proposals discussed above are beset with the following difficulties: (i) the amount of converted energy is small; (ii) slow response;

Fig. 4.8. An example of a mechanochemical engine with a collagen (crosslinked with formaldehyde) driving belt. (*a*) Diagram of the device. A–B distance: 20.0 cm, $R_2 = 3.44$ cm, $r' = 3.05$ cm, $R_1 = 3.55$ cm, $r = 3.15$ cm, [LiBr] = 11.25 M and 0.3 M. Then, the system is driven at a rate of 40 r.p.m. without load for 1 h. The maximum output is 0.03 W (g collagen)$^{-1}$. (*b*) Actual engine. (From I. Z. Steinberg, A. Oplatka & A. Katchalsky, *Nature, Lond.*, **210**, 568 (1966).)

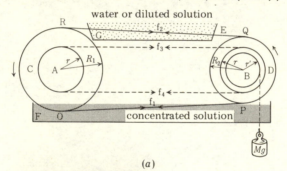

(*a*)

(b)

(iii) the materials employed as energy sources are practically inconceivable as energy resources. In contrast to the apparently topical chemical → mechanical direct energy conversion, the problems of thermal → mechanical energy conversion are less topical than the former but will be more significant in practice. This is because of: (i) utilization of a low temperature heat source – temperature differences of several tens of degrees centigrade will be sufficient; (ii) the process is operative by repetition of heating and cooling involving no chemical reactions – this is a great advantage to simplify the device; (iii) a faster response is expected. Y. Osada made use of the large volume change brought about by the complex formation–dissociation between polyethyleneglycol (PEG) and poly(methacrylic acid) (PMAA), which is sensitive to temperature. (Y. Osada & Y. Saito, *Makromol. Chem.*, **176**, 2761 (1975).) As shown in Fig. 4.9, the size of the PMAA film

Fig. 4.9. Contraction of PMAA film by temperature change. A PMAA film (10 × 23 mm, 4.7 mg in dry state) with a load of 490 mg is dipped in (1) 70 ml of pure water or (2) 70 ml of aqueous polyethylene glycol (mol. wt: 2000, 0.015 unit mol dm^{-3}) solution. (From Y. Osada & H. Saito, *Nippon Kagaku Kaishi*, 173 (1976).)

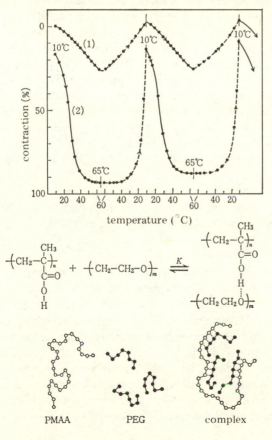

becomes very sensitive to temperature in the presence of a small amount of PEG, and the expansion–contraction indeed reaches 90% of the film length for the temperature difference between 10 and 40 °C. The interaction force between PMMA and PEG is not simple hydrogen bonding. Participation of hydrophobic interaction between the methyl groups of PMAA and the main chain ethylene groups of PEG is postulated. In this temperature region, hydrophobic interaction is enhanced with increasing temperature. Using the same PMAA film complex formation is a function of the concentration and molecular weight of PEG. With increasing molecular weight, complex formation becomes so facilitated that the complex is rendered stable over the temperature range used, and the contraction-expansion phenomena are not observed. The optimum molecular weight of PEG is between 2000 and 3000. When PEG concentration is too low, the mechano-chemical process is not reversible. Under the best conditions, the work by the PMAA film is $\sim 2 \times 10^{-2}$ J (g polymer)$^{-1}$ and the contracting power is $\sim 1\,\mathrm{kg\,cm}^{-2}$.

The stability of complex formation is determined by the balance between hydrogen bonding and hydrophobic interaction. The complex stability of the PAA-PEG system is less than that of the PMAA-PEG system since participation of CH_3 groups in hydrophobic association is not expected. The PMAA-poly (vinyl pyrrolidone) system has a very high stability constant and therefore the temperature sensitivity is less than that of the PMAA-PEG system. In an aqueous alcohol system, hydrophobic interaction in the PMAA-PEG system disappears and the complex stability decreases with increasing temperature. As a consequence, a reversed temperature effect on the size of polymer is observed.

We can think of many tricks with interpolymer complexes. Although operation in a wet system is a drawback, similar to the chemical → mechanical energy conversion systems, an identical device to that shown in Fig. 4.8 would be possible using hot and cold baths instead of salt baths. Either way, the acquisition of mechanical energy in the form of reciprocal movement is inefficient. Such a process may be compared to the prototype steam engine by Newcomen. Operation in a dry system is more favourable, but the control of association and dissociation of polymer molecules in a dry system is difficult.

4.3.3   *Photoenergy → mechanical energy*

Reversible photochemical deformation of polymer material is the principle of photo- → mechanical energy conversion. As photophysicochemical processes, the excited state interactions such as excimer and exciplex formation could be origins of polymer deformation. These phenomena, however, are short-lived ($10^{-7} \sim 10^{-9}$ s) and the high concentration storage under stationary photo-irradiation is not practicable. Consequently, photoisomerization is considered to be a suitable elementary process. When the thermal reverse reaction is slow, a large difference in stationary concentrations in the dark and under photo-irradiation is obtained whereas the recovery by dark reaction requires a longer time. A longer response time is disadvantageous because of slow energy conversion cycles. The photo- and thermal reaction cycle of a solid polymer can be operated in dry state, which is an advantage.

The best known system is based on the molar volume change of spirobenzo-pyran brought about by photoisomerization. The system demonstrated by G. Smets consists of a copolymer of ethylacrylate and 8,8'-bisacryloxymethyl-enebisphotochrome D. (G. Smets, *Pure Appl. Chem.*, **42**, 516 (1975); G. Smets, *J. Polym. Sci., Polym. Chem. Edn*, **13**, 2223 (1975).) The polymer film (length: 41 mm, width: 5.5 mm, thickness: 0.48 mm) with a load of 26.8 g contracts by 1.5% under photoirradiation. Immediate recovery of size is observed when photoirradiation is terminated. This process can be repeated many times.

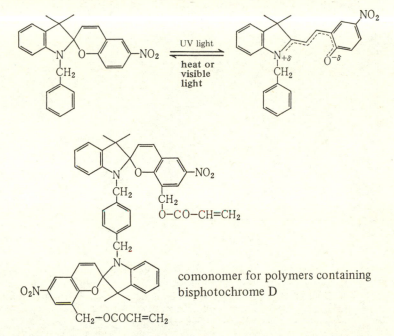

comonomer for polymers containing bisphotochrome D

Photoisomerization is strongly influenced by matrix properties. The fading reaction (thermal back reaction) in non-polar polystyrene matrix is faster than that in polar methyl methacrylate matrix by a factor of 10. Isomerization as well as many other reactions accompanying molecular motion and volume change depends on the rigidity of the matrix. The fading rate and its tempera-ture dependence show different trends when photochrome C (Fig. 4.10) blended with a polyester is compared with that chemically bound to the polyester. In particular, the chemically bound system exhibits a sharper temperature depen-dence around T_g (glass transition temperature). The photocontraction–thermal expansion profiles are shown in Fig. 4.11 for a copolymer of ethyl acrylate with photochrome C (R = methacrylate). There is an optimum load for the maximum photocontraction.

Photoisomerization of diazobenzene can also be applied for photocontractive polymers. As indicated by self association phenomena of azo dyes in solutions, the *trans*-form associates more easily. When diazobenzene is bonded to a polyelectrolyte, the attraction by hydrophobic association of diazobenzene and the Coulombic repulsion are equilibrated to give a certain polymer chain dimen-

Fig. 4.10. Temperature-dependent fading-rate constant ($\ln k$) of photochrome C built in a polyester (curve a) and photochrome C propionate doped in poly(bisphenol A pimelate) matrix. (From G. Smets, *Pure Appl. Chem.*, **42**, 516 (1975).)

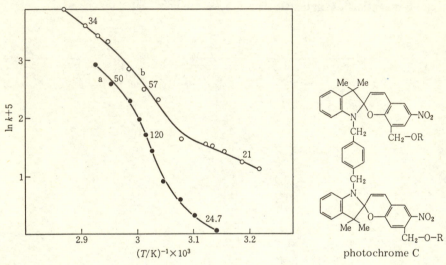

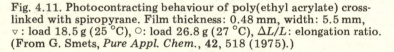

photochrome C

Fig. 4.11. Photocontracting behaviour of poly(ethyl acrylate) crosslinked with spiropyrane. Film thickness: 0.48 mm, width: 5.5 mm, $\triangledown$: load 18.5 g (25 °C), $\circ$: load 26.8 g (27 °C), $\Delta L/L$: elongation ratio. (From G. Smets, *Pure Appl. Chem.*, **42**, 518 (1975).)

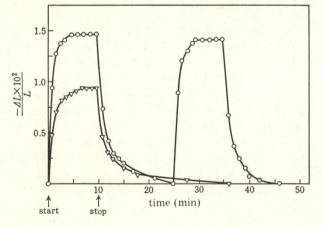

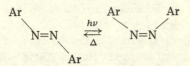

sion. By photoirradiation, the *cis*-form which is sterically unfavourable for molecular association increases and the polymer chain expands. Another method is based on the fact that the *cis*-form of charged azostilbene dye (crysophenine) is hardly adsorbed on to the polymer chain. As shown in Fig. 4.12, photo-

Fig. 4.12. Change of molecular interactions due to photoisomerization. (*a*) The balance between hydrophobic interaction and Coulombic interaction changes (the negative charges: $-COO^{\ominus}$, etc.). (*b*) Change of dye absorption to polymer due to isomerization of the azostilbene dye stuff. (From R. Lovrien, *Proc. Natl. Acad. Sci., USA*, **57**, 237 (1967).)

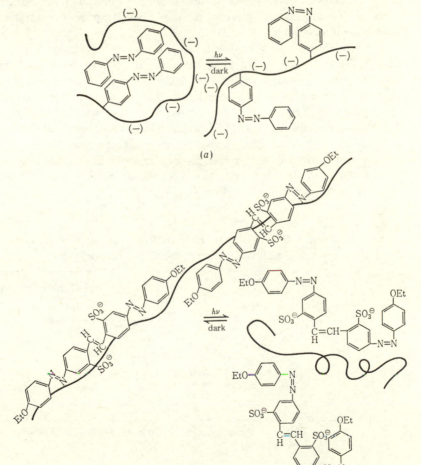

irradiation alters the solution viscosity. In particular, the phenomenon is prominent in polymer-adsorbed dye systems.

Intrinsic reactivity of a functional group is certainly important as a factor determining the energy converting ability of polymers. However, small differences in polymer structure influence very much the thermal (recovery) reaction rate, as presented in Table 4.4. An apparently curious phenomenon is observed: that the recovery is even faster in polymer film than in solution for copolymer (b). Since specific polymer materials are mostly used as solid materials, the functionality–polymer structure correlation should be well understood as a basis of materials design. The examples presented in this section suggest that the energy conversion functions stem from intra- and interpolymer interactions between functional groups. Molecular interactions in polymers are of course controlled by primary (chemical), secondary and higher order structures including differences in crystallinity and molecular orientation. The new methodology is gradually coming up to elucidate polymer functions as molecular phenomena relevant to molecular interactions.

No attempt has been made to convert photoenergy to rotational mechanical energy. Referring to Fig. 4.8, such a device could be easily constructed. Since operation in a dry state is possible, the device could be simpler. Because there is no material deterioration, the workability is thought to be better than the machine in Fig. 4.8, which requires exchange of salt solution. The upper part of the belt would be irradiated by sunshine whereas the lower part would be covered up to allow the dark reaction. Furthermore, if a black body having a large heat absorption were used as the awning, the dark reaction would be accelerated at elevated temperatures and therefore solar energy could be efficiently used over a wide wavelength region.

Table 4.4. *Rate of thermal cis → trans isomerization of styrene-vinyl-N,N-dimethylaminoazobenzene (9:1) copolymers*

	Half life of recovery (s)	
Copolymer structure	Solution	Film
(a) $\text{(P)}\!-\!\!\bigcirc\!\!-N{=}N{-}\!\bigcirc\!\!-N(CH_3)_2$	30	150
(b) $\bigcirc\!\!-N{=}N{-}\!\bigcirc\!\!-N(CH_3)_2$ (P)	15	7

Source: From H. Kamogawa, M. Kato & H. Sugiyama, *J. Polym. Sci.*, A-1, **6**, 2967 (1968).

4.4   **Photoenergy → Chemical energy conversion**
       The reason why the photosynthetic system is unrivalled as an energy
conversion device is that it leads to the production of organic compounds
(foods) from water and carbon dioxide. From the simple viewpoint we can
think of photodimerization and many others as examples of endothermic
photoreactions, but let us consider the possibility of mimicking photosynthesis
and the application of polymers.

4.4.1   *Simulation of the photosynthetic system*
       As discussed in §4.2, the functions of the photosynthetic system are
divided as follows. (i) Photoabsorption by antenna pigments and energy collection by energy transfer to the reaction centre. (ii) Excitation of the reaction
centre to drive electron transfer processes. (iii) Coupling of oxidation of water
and reduction of carbon dioxide. (iv) Separation of oxidation and reduction
products by a membrane. Firstly, aside from polymer application, we will
extract the processes for which partial simulation has succeeded, and then
discuss possible modifications using polymers. A major difference from what
was described in §4.3 is that here the mechanical properties of polymers are
of auxiliary importance. Consequently, it will be a promising strategy to combine the information relevant to (i)–(iv) with the excellent mechanical properties
of polymer. This approach will lead to the total energy conversion system
including energy harvesting as well as product separation.
       For function (ii), many sensitizers entirely different from chlorophyll have
been recently found to drive electron transfer cycles *in vitro*. The reactions are
called 'electron transfer (or transport) sensitization'. Electron transfer via photoexcitation or photoionization is not uncommon. Since the first example by
G. N. Lewis (1942) of photoionization of N,N,N',N'-tetramethyl-p-phenylene-
diamine, there have been a number of publications. The phenomenon has not
however been looked at as a method of sensitization. Until recently, sensitized
reactions in photochemistry have meant exciting the reacting molecules to the
triplet state by means of triplet energy transfer from the excited sensitizer.
Although singlet-state sensitization via singlet energy transfer from the singlet
excited state of sensitizer is possible, the absorption region of the sensitizer has
to be on the blue of that of the energy accepting species. Namely, singlet
sensitization does not meet the general requirement of expanding the effective
wavelength towards red.
       The triplet sensitizers are mostly carbonyl compounds having a narrow energy
gap between singlet (S) and triplet (T) states so that triplet sensitization of
a molecule having a large S–T energy gap is achieved by a photoenergy which is
lower than that necessary to excite the molecule directly to the S state. This is
the spectral sensitization shown in Fig. 4.13. Electron transfer sensitization is
the application of enhanced electron affinity and decreased ionization potential

in the excited state and consequently of facilitated electron transfer reactions to couple more than two electron transfer cycles. In other words, sensitizer (S) acts as a photo-driven pump to pump out electrons from D to A. This is a prototype of the reaction centre although the features of the reactions are not at all comparable with photosynthesis.

For function (iii), coupling of reduction of carbon dioxide with oxidation of water has been demonstrated *in vitro*. (S. Tazuke & N. Kitamura, *Nature, Lond.* **275**, 301 (1978).) This is a good example of borrowing Nature's wisdom.

Fig. 4.13. Electron transfer sensitization (*a*) and energy transfer sensitization via either the triplet (*b*) or singlet (*c*) state.

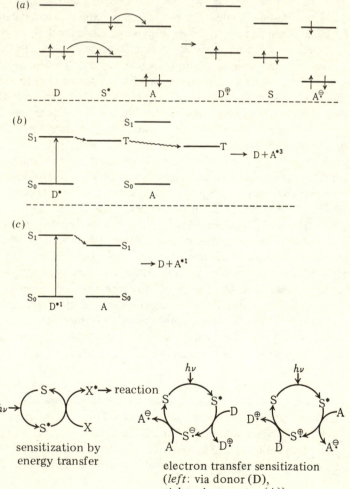

sensitization by
energy transfer

electron transfer sensitization
(*left*: via donor (D),
right: via acceptor (A))

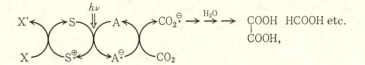

S: aromatic hydrocarbons such as pyrene

A: acceptors such as 1, 4-dicyanobenzene and 9, 10-dicyanoanthracene

X: not clear, but $OH^{\ominus} \rightarrow \frac{1}{2}H_2O_2 + e$ (oxidative decomposition of water) is included

In photosynthesis, photoenergy stored in the form of NADPH and ATP is used to conduct reductive fixation of CO_2 via a series of dark reactions. Considering the photosynthetic system simply as a black box which produces organic compounds photochemically from CO_2 and water, the technical breakthrough is already within our reach. At the moment, the effective wavelength is at the maximum 450 nm, which is much shorter than that of photosynthesis. The products are not elegant and edible organic compounds. Furthermore, the CO_2 uptake rate is much less than 100 μl (mg sensitizer)$^{-1}$ h^{-1} with a 300 W high pressure mercury lamp whereas chlorella (*Chlorella pyrenoidosa*) takes up CO_2 at a rate of 3000 μl (mg chlorophyll)$^{-1}$ h^{-1}. Nevertheless it is no longer a dream to store photoenergy from CO_2 and water by endothermic photoreactions. Recent advances in photofixation of CO_2 and water photolysis are very rapid. The use of inorganic semiconductors in reducing CO_2 to CH_4 has been reported. (T. Inoue, A. Fujishima, S. Konishi & H. Honda, *Nature, Lond.* **277**, 637 (1979).) Semiconductors can conduct simultaneous multiple electron transfer so that the multistep reduction of CO_2 is facilitated.

The reasons for the low quantum efficiency are the lack of antenna chlorophyll equivalents, the ease of back electron transfer ($S^{\oplus} + A^{\ominus} \rightarrow S + A$) because of homogeneous reactions and, in addition, the existence of oxidized products (O_2, H_2O_2, etc.) and reduced products (HCOOH, $(COOH)_2$, etc.) as a mixture. This situation will induce secondary decomposition of these products resulting in a decreased quantum efficiency. A procedure to improve the efficiency is to utilize a Coulombic field to separate $\oplus$ and $\ominus$; for example, when an anion radical is generated in a cationic micelle, the cation radical pairing with the anion radical is repelled from the micelle due to repulsion between $\oplus$ charges, and the remaining anion radical is stabilized in the micelle. As a consequence, the back electron transfer is retarded and the lifetime of the anion radical is greatly extended. (B. Katsušin-Ražem, M. Wong & J. K. Thomas, *J. Amer. Chem. Soc.*, **100**, 1679 (1978).) The use of micellar or polymeric systems is not essential. A small molecular redox sensitizer bearing an ionic group can achieve a 100% charge separation as well. (S. Tazuke, Y. Kawasaki, N. Kitamura & T. Inoue, *Chem. Lett.*, 251 (1980).)

An energy-collecting device comparable to antenna pigments has not been constructed. Although the method to arrange various pigments having slightly different energy levels from each other in the order of their energy levels has not been established, there are attempts to increase photoabsorption cross section by means of energy migration among chromophores of identical energy levels. The simplest method to make chromophore aggregates is to use polymers. Energy migration among chromophores attached to a polymer is a common phenomenon, as observed in many polymers bearing aromatic groups such as the phenyl groups in polystyrene and carbazole groups in poly(9-vinylcarbazole). When D groups are incorporated in a polymer, the reaction $D* + A \rightarrow D + A*$ or $D^{\oplus} + A^{\ominus}$ can be induced by photoenergy absorbed at somewhere apart from the site of A as shown below.

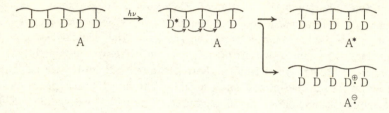

This is, however, due to random energy migration, and the directional energy transfer from high energy sites to low energy sites is not expected. The efficiency is consequently not high. Nevertheless, the efficiency is better than that of the isolated D in which D can encounter A only by diffusion. Being a polymer, on the other hand, a negative factor of reduced diffusion constant has to be taken into account. The positive factor of increasing reaction diameter due to energy migration is in part cancelled by the negative factor of reducing the travelling diameter of D within its lifetime due to the decreased diffusion constant. When sensitizers are fixed in a membrane, like the photosynthetic system, and translational diffusion is unfavourable, polymeric sensitizers will be advantageous in a relative scale.

Energy transfer (migration) is either between singlet states or between triplet states. The phenomenological difference between these two types is the effective distance of energy transfer. Singlet energy transfer proceeds via dipole–dipole interactions without overlap of electronic orbitals (Förster mechanism), and therefore it is possible over a long distance ($10^0 \sim 10^1$ nm). Since triplet energy transfer requires electron exchange interactions, and consequently overlap of electron clouds, the effective distance is $< 10^0$ nm. However, the close approach of an energy transferring pair is not always favourable. If the energy transferring pair forms an excited complex (excimer or exciplex) and is stabilized, further energy transfer to other chromophores is prevented. Also, the mutual arrangement of D and A (orientation factor) influences the energy transfer efficiency. The orientation factor is averaged in homogeneous systems whereas the optimum arrangements of chromophores should be possible in the future by means of polymer-bound sensitizers. The secondary and higher structures such as helix formation and orientation by stretching are important characteristics of polymer systems.

When anthracene molecules are attached to a polymer chain with a regular interval but without considering their orientation, singlet energy can migrate between two anthracene molecules separated by 10~20 atoms. Using the polymer as sensitizer, the photosensitizer efficiency of the polymer is higher than that of the relevant low molecular weight model in the photooxidation of leuco crystal violet (LCV) (Fig. 4.6b). (S. Tazuke, H. Tomono, N. Kitamura & N. Hayashi, *Chem. Lett.*, 85 (1979).)

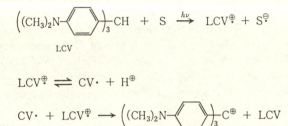

Providing that molecular aggregates of photosensitizer are effective, it is not necessary to stick on covalently bonded polymers. Non-covalently bonded systems such as micelles or liquid crystals are also expected to sustain the energy migration effect. The constituent molecules stay in a micelle for 10^{-2}~10^{-5} s, indicating their rapid exchange, although the lifetime of the micelle itself may be long. Either using functional groups bound to surfactant molecules or solubilizing functional molecules in micelles, sustaining molecular orientation of functional groups is difficult.

Separation of oxidation and reduction sites – function (iv) – is the most difficult part. As we have already seen, micellar systems are capable of redox site separation. So are bilayer membrane systems. It is now possible to demonstrate redox coupling between the inside and the outside of a bilayer membrane. However, it will not be possible to take out the redox products separately and continuously. For the purpose of macroscopic material separation, a charge-separation membrane having reasonable mechanical strength is required. Comparing the principles of electron transfer sensitization with the thylakoid membrane (Fig. 4.7), mimicry may be possible by constructing a membrane with an S that permits efficient flow of electrons. Although higher order functionalities such as proton transport are required as the second stage, discussions are confined to electron transport. The first candidates for the material are conductive polymers represented by polyacetylene.

Semiconductive polymers are divided into two categories: one is photoconductive polymers such as poly(9-vinylcarbazole), and the other is highly conductive material ultimately aiming at superconductors. The former are already in use as image-recording materials (electrophotography). Photoconductive polymers have the following characteristics. (i) Low dark conductivity; (ii) because of (i) the material is suitable by way of its bulk properties for use as semiconductors but not for the formation of p–n junctions; (iii) good processability and easy formulation; (iv) good transparency. These properties are useful for electrophotography in which a surface-charged photoconductive plate is subjected to imagewise irradiation to form an electrostatic image. On the other hand they are not adequate for photocells or rectifiers because of high internal resistance. High conductivity cannot be expected from weak interactions between aromatic side chains in vinyl polymers bringing about some carrier mobility via hopping of electrons or holes. Consequently, fundamentally different approaches of molecular designing are needed for photoconductors and highly conductive materials.

For many years there have been attempts to synthesize highly conductive organic compounds, such as electron donor–acceptor complexes (EDA complex) and ion-radical salts, and to make polymer materials from them. By far the most

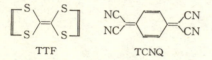

TTF   TCNQ

stimulating achievement was the report by A. J. Heeger and his coworkers (1973) on the conductivity of tetrathiafluvarene (TTF)–tetracyanoquinodimethane (TCNQ) single crystals. (L. B. Coleman, M. J. Cohen, D. J. Sandman, F. G. Yamagishi, A. F. Garito & A. J. Heeger, *Solid State Commun.*, **12**, 112 (1973).) Although they once enjoyed the reputation of being organic super conductors, their 'super conductivity' was eventually denounced. However, the crystals are indeed highly conductive and the results of X-ray crystallography indicate that both TTF and TCNQ form segregated column-like stacks. For easy use as a functional material, all attempts to make polymers containing TTF and/or TCNQ resulted in reduced conductivity since the stacks of TTF and/or TCNQ were destroyed owing to the randomness of the polymer structure. $(SN)_x$ is a polymer crystal prepared by photo- or thermal solid-state polymerization. Although the conductivity is high, the processability of $(SN)_x$ is poor. Polymerization of acetylene into sheet by H. Shirakawa and his coworkers (1977) is a genuine way to compromise conductivity and processability. (H. Shirakawa, E. J. Louis, A. G. MacDiarmid, C. K. Chiang & A. J. Heeger, *Chem. Commun.*, 578 (1977).) When acetylene gas is introduced into a reaction vessel whose inside wall is coated with Zeigler–Natta catalyst, a film-like polyacetylene with metallic lustre is obtained. In Table 4.5, the conductive properties of polyacetylene with various dopants are shown.

The reasons why attention is focussed on polyacetylene are the following. (i) A polymer film having conductivity as high as $10^3 \, \Omega^{-1} \, cm^{-1}$ has been achieved for the first time. (ii) By the technique of moulding polymerization, a flexible polymer film of the desired size and thickness can be prepared. The film can be stretched to three times its original length. After stretching, an anisotropic conductivity of $\sigma_\| / \sigma_\perp \approx 15$ is observed, where $\sigma_\|$ and $\sigma_\perp$ are the conductivities along and perpendicular to the direction of stretching, respectively. The anisotropy would suggest that carrier transport occurs along the $+CH=CH\rightarrow_n$ main chain. (iii) The mode of conduction is p-type with electron acceptors such as halogens and AsF_5 as dopants, whereas the addition of electron donors such as $Na^\oplus$ (naphthalene)$^\ominus$ gives rise to n-type conduction. Furthermore, p–n junction formation and a rectifying effect are observed by pressing p-type and n-type polyacetylenes together. (iv) By controlling doping species and its level, a variety of conductivity can be chosen between 10^3 and $10^{-9} \, \Omega^{-1} \, cm^{-1}$. (v) The conductivity per unit weight is comparable to mercury (Table 4.6). Comparing $(SN)_x$ and $(TTF)(TCNQ)$ in polycrystalline form, the composition of *cis*-[CH-$(AsF_5)_{0.14}]_x$ exhibits the highest σ.

Intrapolymer conduction of all-conjugated polymers should be excellent. The powdered samples would show insufficient conduction between polymers in comparison with the film sample and consequently, the conductivity is small. The famous model of W. A. Little assumes infinite repetition of the repeating unit. (W. A. Little, *Phys. Rev.*, **134**, A1416 (1964).) In practice, the carrier mobility between molecules of finite molecular weight has to be considered. Better orientation (namely, close to a single crystal) is an important factor.

Table 4.5. *Conductivity (σ) of various polyacetylenes* $+CH\rightarrow_x$

Substance	$\sigma \, (\Omega^{-1} \, cm^{-1})$ at 25 °C
cis-$(CH)_x$	1.7×10^{-9}
trans-$(CH)_x$	4.4×10^{-5}
trans-[$CH(HBr)_{0.04}]_x$	7×10^{-4}
trans-$(CHCl_{0.02})_x$	1×10^{-4}
trans-$(CHBr_{0.05})_x$	5×10^{-1}
cis-$(CHI_{0.25})_x$	3.6×10^{2}
trans-$(CHI_{0.22})_x$	3.0×10^{1}
trans-[$CH(AsF_5)_{0.03}]_x$	7×10^{1}
trans-[$CH(AsF_5)_{0.10}]_x$	4.0×10^{2}
cis-[$CH(AsF_5)_{0.14}]_x$	5.6×10^{2}
trans-[$Na_{0.28}(CH)]_x$	8×10^{1}

Source: C. K. Chiang, M. A. Drug, S. C. Gau, A. J. Heeger, E. J. Louis, A. G. MacDiarmid, Y. W. Park & H. Shirakawa, *J. Am. Chem. Soc.*, **100**, 1014 (1978).

Polyacetylene, as a π-conjugated system behaving both as D and A, resembles aromatic hydrocarbons. If an electron transfer sensitizer system could be constructed by the use of this membrane, the oxidation and the reduction sites would be separated as shown in Fig. 4.14 and the reaction system would be of practical interest. The well-matched absorption spectra of polyacetylene with solar spectra is also an advantage. Although the carrier mobility has not been determined, the value of $560 \, \Omega^{-1} \, cm^{-1}$ for *cis*-$[CH(AsF_5)_{0.14}]_x$ provides a carrier mobility (μ) of $1 \, cm^2 \, V^{-1} \, s^{-1}$, assuming one carrier per one AsF_5 molecule. The figure is comparable with that of crystalline anthracene, and much larger than that of poly(9-vinylcarbazole) ($10^{-6} \sim 10^{-9} \, cm^2 \, V^{-1} s^{-1}$). It might therefore function as a electron transport membrane even if the film is thick enough (of the order of micrometres) to impart sufficient mechanical strength. The problem of maintaining charge balance between electron-releasing and electron-accepting sides remains to be solved. Ion transport coupled with electron transport across the membrane is required. Recent evidence shows that the rate of electron transport in the bilayer membrane is controlled by the rate of potassium ion permeation across the membrane. (M. Calvin, I. Willner, G. Laane & J. W. Otvos, *J. Photochem.*, **17**, 195 (1981).)

4.4.2 *Polymer-supported catalyst*

In contrast to the ambitious future story of mimicking photosynthesis, the solar heating problem is being looked at from a more practical standpoint. In a solar-heated house, the heat medium is warmed up by solar energy and stored underground. Together with the physical process, there are attempts to store energy by means of endothermic photoisomerization reactions. The conditions for utilizing the reaction,

Table 4.6. *Conductivity per unit weight*

Material	Density d (g cm^{-3})	σ (Ω^{-1} cm^{-1})	σ, after correcting density σ/d (cm^2 Ω^{-1} g^{-1})
Cu	8.92	5.8×10^5	6.5×10^4
Au	19.3	4.1×10^5	2.1×10^4
$(SNBr_{0.4})_x$ (single crystal)	2.67	3.8×10^4	1.4×10^4
Fe	7.86	1.0×10^5	1.3×10^4
$(SN)_x$ (single crystal)	2.30	3.7×10^3	1.6×10^3
Hg	13.6	1.0×10^4	7.4×10^2
cis-$[CH(AsF_5)_{0.14}]_x$ (polycrystalline film)	0.8	5.6×10^2	7.0×10^2

Source: C. K. Chiang, M. A. Drug, S. C. Gau, A. J. Heeger, E. J. Louis, A. G. MacDiarmid, Y. W. Park & H. Shirakawa, *J. Am. Chem. Soc.*, **100**, 1015 (1978).

$$A \underset{\Delta}{\overset{h\nu}{\rightleftharpoons}} B,$$

are: (i) large absorbance of A and its matching with the solar spectrum; (ii) high quantum efficiency of the A → B conversion; (iii) discrete absorption bands for A and B; (iv) a large and positive enthalpy change for the A → B reaction; (v) stability of B at room temperature and facile recovery of B from A either thermally or catalytically; (vi) low cost of A. Among proposed compounds, the mutual conversion of derivatives of norbornadien (NBD) ⇌ quadricyclene (Q) seems to be most promising. As shown in Table 4.7, the energy storage efficiency of $1.2\,\mathrm{J\,g^{-1}}$ is not sufficient. Nevertheless, this figure is still better than those for

Fig. 4.14. Proposed electron transport membranes to separate oxidizing and reducing sites. (a) a model of a thylakoid membrane with a phospholipid bilayer. (From M. Calvin, *Acc. Chem. Res.*, 11, 369 (1978). (b) A system with a semiconductive polymer membrane.

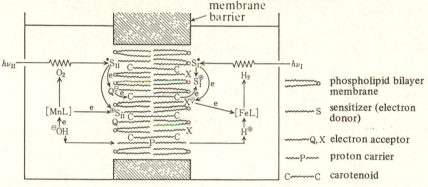

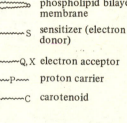

(a)

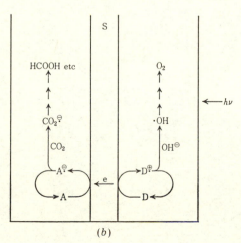

(b)

the rest of the compounds. The low price plus the ease of circulating NBD and subsequently formed Q, both being liquid at room temperature, are advantageous.

Polymers are not energy-storing materials by themselves but are useful as catalysts. Photochemical conversion of NBD to Q is sensitized by CuCl, other low valency metal complexes, or triplet sensitizers. Also the reverse thermal process is catalysed by Pd(NBD)Cl, [Rh(NBD)Cl]$_2$, and others. If these sensitizers and catalysts are bound to polymers, the expected merits are as follows. (i) Since the sensitizer can be fastened in the irradiating compartment, mixing of the photosensitizer with the heat-generating catalyst is avoided, as shown in Fig. 4.15. (ii) The local concentration of NBD in the periphery of photosensitizer will be increased because of the good affinity of the non-polar polymer chain to NBD. (iii) When polymeric sensitizer has the energy-collecting (antenna) function, the sensitizer efficiency will be enhanced.

The following polymer-bound sensitizers based on porous polystyrene have been proposed. (R. R. Hautala, J. Little & E. Sweet, *Solar Energy*, **19**, 503 (1977).) The problem of absorbing the shorter wavelength alone relative to

Table 4.7. *Energy storage by endothermic photoreaction*

Photoreaction	Longest wavelength of effective irradiation (mm)	Quantum yield	ΔH (J g^{-1})
$NOCl \longrightarrow NO + \frac{1}{2}Cl_2$	600	1.0	575
NBD $\longrightarrow$ Q R=H	366	0.5	1200
R=CO$_2$CH$_3$	334	0.5	374
(Ph, Ph naphthalene cyclobutene photoreaction)	366	0.3	400
(dicyclopentadienone photoreaction)	366	0.4	470
(N–CO$_2$C$_2$H$_5$ azepine photoreaction)	458	0.013	255

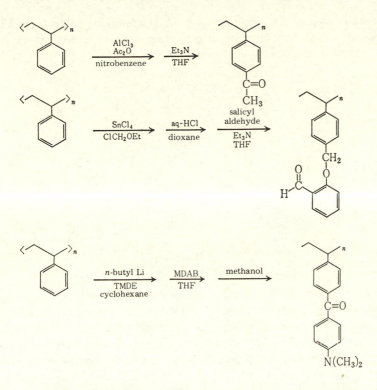

Fig. 4.15. A conversion system for photo- → chemical → thermal energy using polymer-supported catalysts. (From R. R. Hautala, J. Little & E. Sweet, *Solar Energy*, 19, 504 (1977).)

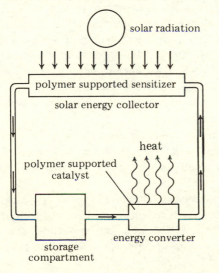

whole solar spectrum has remained unsolved. The following polymer-bound heat-generating catalysts have been reported to promote the $Q \rightarrow NBD$ recovery. (R. B. King & E. W. Sweet, *J. Org. Chem.* **44**, 385 (1979).)

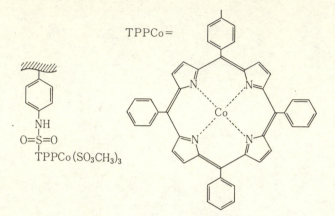

Research along this line should be practical and aimed at specific objectives, taking the cost–performance relation into account. According to a trial calculation in the United States (G. Jones, II, T. E. Reinhardt & W. R. Bergmark, *Solar Energy*, **20**, 241 (1978)) the construction cost should be around or less than 2000 dollars when a cycle of the energy-storing process absorbs and releases 10^6 kJ and the service time is 3 years (or 300 cycles). Considering the efficiency, the cost of energy-storage material is calculated as $40 \sim 160$ yen kg^{-1}. Although the idea of polymer supported catalyst is of interest the technological success will lie in the durability of the catalyst over several years.

4.5 Polymers in photoenergy → electrical energy conversion

When a photoinduced charge separation is drawn off as a potential difference, this constitutes a photocell. Both dry and wet photocells are available. The former represented by silicon photocells is the application of semiconductors. The silicon photocell is already in practical use in cases where the cost can be disregarded, such as the for power supply in an artificial satellite. However, as the silicon single crystal is the key material of the dry photocell, the production cost remains prohibitively high for wide use.

In addition, it is a contradiction that the production of the photocell itself is an energy-consuming industry. Now, research aiming at the application of amorphous silicon whose production processes are simpler relative to the single crystal is in progress.

Let us consider the possibility of conductive polymers such as polyacetylene. As mentioned in the previous section, polyacetylene satisfies most conditions as a low-resistance semiconductor, and furthermore, can be obtained as a film. Although there are the unsolved problems of forming stable p–n junctions and preventing material deterioration, the polymer is the most promising among presently available polymer materials. Previously, the application of the photo-electromotive force generated by the poly(9-vinylcarbazole)-2,4,7-trinitro-fluorenone system to photocells, and the total conversion efficiency, were discussed. (P. J. Reucroft, T. Takahashi, H. Ullal, *Appl. Phys. Lett.*, 25, 644 (1974).) Under the strong electric field of $10^6 \sim 10^8$ V cm^{-1}, the efficiency may reach 1~2% at the maximum. To apply a strong electric field, however, the film thickness has to be reduced and therefore the photoabsorbance decreases. At a film thickness of 1 μm, the upper limit of carrier generation efficiency is around 10^{-3} and the overall efficiency is not greater than 0.01%. Such highly resistant semiconductive polymers cannot be a substitute for low-resistance inorganic semiconductors.

Since the formation of p–n junctions with polyacetylene is possible, this polymer can be used in photocells, in principle. The low resistivity of poly-acetylene is a favourable factor to reduce the internal resistance of photocells. However, the competing technology aimed at using amorphous silicon as a material for planar photocells is progressing rapidly. Stabilization of the amorphous state and improvement of efficiency have been achieved to some extent. The energy conversion efficiency is now approaching ~10% for flexible planar composition. Probably, no organic polymer could compete with inorganic semiconductors in this field at present, and is unlikely to do so in the near future. Inorganic polymers may open a new field, but that is another story.

Wet photocells are inferior to dry photocells with respect to the output per unit volume and unit weight, and service time. However, photochemical batteries will be useful to store photoenergy and to use it when needed. In the photo-synthetic system, the charge-separating capability of the excited thylakoid membrane may be converted to electricity by means of oxidation at the anode and reduction at the cathode, and a photocell will be constructed. The bilayer membrane in Fig. 4.14(*a*) is mechanically not strong enough and has to be supported on a millipore filter. Here again strong polymeric membranes for charge separation are required. Similar to photosynthesis mimicry, the back-electron transfer in the initial stage of charge separation has to be suppressed. A set of synchronized electron transfer would be a solution to separate opposite charges.

The photoredox combination of thiazine dye D (thionine, methylene blue, etc.) with Fe^{2+} ion is a well-known wet photocell. A shortcoming is the fast back reaction, $Fe^{3+} + DH_2$, in the dark. If the back reaction is inhibited, energy

$$\frac{1}{2}D + H^+ + Fe^{2+} \xrightarrow{h\nu} \frac{1}{2}DH_2 + Fe^{3+}$$

$$\text{anode}: \frac{1}{2}DH_2 \longrightarrow \frac{1}{2}D + H^+ + e$$

$$\text{cathode}: Fe^{3+} + e \longrightarrow Fe^{2+}$$

storage becomes possible. Instead of the $Fe^{3\oplus}/Fe^{2\oplus}$ pair, if ascorbic acid or allylthiourea (Red) is used with azurene (Az) as an electron transfer mediator, the separation between $D^{\ominus}$ and $Red^{\oplus}$ is accomplished. (M. Hafner, U. Steiner, S. Schreiner & H. E. A. Kramer, *Z. Phys. Chem.*, *N. F.*, **86**, 220 (1973).) The back reaction is not accelerated by Az so that the efficiency increases.

$$D^* + Az \longrightarrow \overset{\ominus}{D} \cdot + \overset{\oplus}{Az} \cdot$$
$$\overset{\oplus}{Az} \cdot + Red \longrightarrow Az + \overset{\oplus}{Red} \cdot$$

If one could proceed a little further to separate D and Red in anodic and cathodic compartments, respectively, by a conductive polymer membrane, the photocell would function as a secondary battery capable of repeated charging and discharging.

4.6   Future prospects

There is no practical energy-converting system at the moment in which synthetic polymers play the major role. It should be remembered that all energies from the sun are converted and stored by biological systems to support our human life. From now on should we leave all functions relevant to energy conversion to biopolymer systems? Can synthetic polymers find any room to contribute? I believe that there are many items to which our artificial systems can contribute since the biological systems are not at all ideal as far as an energy conversion system is concerned. The merit, and at the same time the demerit, of a biological system is its self-complete function as a system. Although each component molecule is not highly stable, self-repairing and self-reproduction systems are included in a biological system so that continuous energy conversion is made possible. Furthermore, there is the feed back mechanism to adjust the system to the environments. On the other hand, this beautiful system seems to be maintained with the sacrifice of energy-conversion efficiency. For example, the photosynthetic system has a large light-harvesting part in order to sustain its function even under dim light. Under bright sunshine, it is obviously more efficient if the number of chlorophyll reaction centres is increased whereas that of antenna pigments is reduced. For biological systems, energy conversion is a means of maintaining the species and not an aim in itself. Consequently, biological systems are not ideally constructed for the mono-functional purpose of energy conversion.

For efficient energy conversion, the self-reproduction function and controlled photosynthesis are not necessary. The highest possible efficiency can be obtained by not spending that part of the energy for growth and reproduction that organisms do. Although we can draw many suggestions from Nature, we need not mimic it precisely. If we confine our discussion to energy conversion efficiency alone, a photocell functions far better than photosynthesis by plants, and this is encouraging to our effort to develop artificial photosynthetic systems.

The problem of solar energy chemical conversion is a current topic of interest to chemists. Polymer materials will certainly find uses in the system as structural materials and, it is hoped, as functional materials. However, we should bear in mind that the use of polymers is only one of many possibilities. We should carefully evaluate the applicability of polymers in energy conversion devices in comparison with all alternatives.

This is a targeted and therefore, function-oriented research. We must carefully judge the functionalities of polymer materials. What are the differences of 'polymer effects' supporting the speciality of polymers in comparison with other molecular aggregate systems such as micelles, bilayer lipid membrane, and vesicles? What are the differences of a speciality polymer from a composite material consisting of a physical mixture of functional molecules in a polymer matrix? If we find any merits of polymers through these comparisons, then polymer science can claim its potentiality. The uniqueness of polymer materials lies in the fact that highly functionalized molecules themselves can be used as structural materials. As discussed in §§4.3–4.5, all applications originate from a combination of functions plus mechanical strength. These applications are still in their infancy. Nevertheless, energy conversion is a promising field of research in which polymer science can demonstrate its potentiality as a materials science.

5

TRANSPORT PHENOMENA AND FUNCTIONAL POLYMERS

MANABU SENŌ

5.1 Introduction

The role of transportation is unexpectedly large in biological systems as well as in our industrial and economic systems. This can clearly be seen from the remarkable development of aircraft, motor cars and roads; also in conveyance operations in factories, and metabolic and circulating actions in our bodies. Thus transportation plays a vital part in our lives. Until now, however, the importance of transportation seems to have been neglected in chemistry. So it is of great significance to develop the chemistry of transportation. In this respect, major emphasis has been placed on polymeric materials with special transport functions.

Transport functions of polymers are of two kinds: one relates to the transport properties of polymers themselves, and the other relates to their properties as transport media for other molecules. The polymers serve as transport carriers in the former case and they are used in the form of permeable membranes in the latter case. In any transport system, a high energy efficiency and a high selectivity are especially important. Up to now the chemistry of transport has not been systematically investigated and knowledge on transport functions of substances is very limited. Separation and transport processes are treated differently in present-day chemistry and industry; the transport process is considered more of a problem in chemical engineering, whereas attention is devoted to the separation process in chemistry. Nowadays, however, the transport function, with its high selectivity, is becoming important for the development of highly practical and energy-saving chemical processes.

5.2 Polymer membranes as transport media
5.2.1 *Gas permeabilities of polymer membranes*

Polymer films are widely used as packaging materials. For this purpose, the barrier properties are important to protect the contents from their surroundings. In contrast, large and selective gas permeabilities are required for a separation material. The gas permeation mechanism through a membrane is dependent on the pore structure of polymer comprising it. In the cases of dense homogeneous membranes, which have no macropores, permeating molecules migrate through molecular interstices woven by the polymer chains. This permeation process obeys the solution–diffusion mechanism. The Fick diffusion law is expressed as

$$\frac{\partial c_i}{\partial t} = \frac{\partial}{\partial x}\left[D_i(c_i)\frac{\partial c_i}{\partial x}\right] \tag{5.1}$$

When the diffusion coefficient, D_i, is independent of concentration, c_i, by solving this equation under the conditions where $c_i = 0$ at $t = 0$, and $c_i^{II} = c_i^0$ $(x = 0)$, $c_i^{II} = 0$ $(x = l)$ at $t > 0$ for the system shown in Fig. 5.1, the following expression is obtained for the permeating amount $Q(t)$ through the plane at $x = 0$ till the time t,

$$\frac{Q(t)}{lc_i^0} = \frac{D_i^0}{l^2}\,t - \frac{1}{6} - \frac{2}{\pi^2}\sum_{n=1}^{\infty}\frac{(-1)^n}{n^2}\exp\left(-\frac{D_i^0 n^2 \pi^2}{l^2}t\right) \tag{5.2}$$

$Q(t)$ determines a permeation curve. When t is sufficiently large, the expression $Q(t)$ takes a limiting equation as

$$Q(t) = \frac{D_i c_i^0}{l}\left(t - \frac{l^2}{6D_i^0}\right) \tag{5.3}$$

The analytical expression cannot be obtained if the diffusion coefficient D_i is dependent on concentration. In that case, the permeation curve is convex to the t axis. The permeation behaviour which obeys these formulae is of the Fick's type, which rules gas permeation processes through ordinary amorphous polymer membranes.

From the linear portion of the permeation curve, the steady speed q_s is defined as

$$\lim_{t \to \infty} dQ(t)/dt.$$

The steady value of the permeability coefficient is defined as

$$P = q_s l/P_i^0 \tag{5.4}$$

and then it is expressed as

$$P = DS \tag{5.5}$$

where the integral diffusion coefficient, D, and the solubility coefficient, S, are

Fig. 5.1. Gas permeation through a membrane.

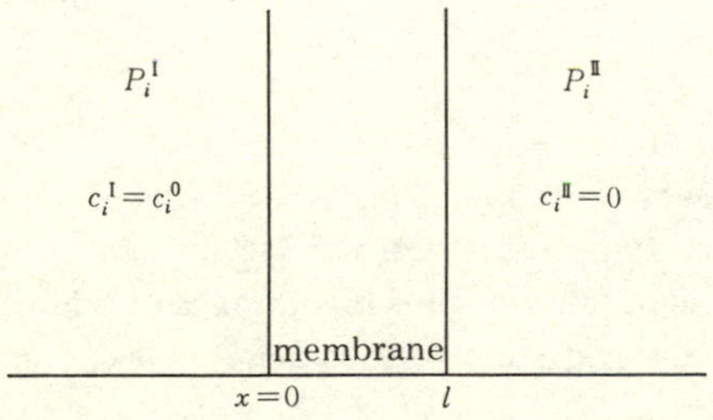

defined as:

$$D = \frac{1}{c_i{}^0} \int_0^{c_i{}^0} D_i(c_i) \, dc_i = \frac{q_s l}{c_i{}^0} \qquad (5.6)$$

and

$$S = c_i{}^0/P_i{}^0 \qquad (5.7)$$

respectively. The permeability coefficient, P, is frequently expressed by using the unit of $cm^3(STP)$ cm cm^{-2} s^{-1} $cmHg^{-1}$. As shown in Table 5.1, the values of P change over rather a wide range from 10^{-8} to 10^{-14} in this unit, but the permeability ratios for two kinds of gases, e.g. P_{O_2}/P_{H_2} and P_{CO_2}/P_{N_2}, remain nearly constant, independent of the nature of membrane material.

For many polymer membranes, the diffusion coefficients of water vapour, as well as those of nitrogen and oxygen gases, show little dependence on concentration. This is observed commonly for the permeation of non-soluble gases through membranes, but the concentration dependence becomes marked as the affinity of a gas to the membrane matrix increases. For instance, the permeation of water vapour is not dependent on the vapour pressure for membranes of polyethylene or poly(vinylidene chloride), slightly dependent at a higher vapour-pressure range for poly(vinyl acetate), and largely dependent for acetyl cellulose and poly(vinyl alcohol). The permeation of water through poly(dimethyl siloxane) and poly(methyl methacrylate) membranes decreases as vapour pressure increases. This non-Fickian behaviour is due to the immobilization of water molecules in the membrane matrix.

It is important to achieve selective permeability for a given gaseous substance through the membrane in order to utilize the membrane as a tool for purification, and separation method for gaseous mixtures. For example, an SO_2-selective membrane is obtained by soaking tetramethylene-sulphone or dihydrothiophene-1,1-dioxide. Concentrating oxygen from air by using a poly(dimethyl siloxane) membrane has been known for a long time, but its selectivity towards oxygen permeation is not especially high, although the permeation rate is very large. A membrane for which the selectivity ratio P_{O_2}/P_{CO_2} is nearly unity is required for substituted artificial lungs; however, in general the permeation rate of CO_2 is larger than that of O_2 because of the higher solubility of CO_2. It was reported that a porous membrane of mean pore size of about $0.5\,\mu m$ exhibits selectivity for O_2 against CO_2 together with an enhanced O_2 permeability compared to a dense membrane.

5.2.2 *Preparation of polymer membranes*

The pore structure of membranes is designated in terms of pore size and distribution, and its pore density or interstitial volume. In terms of pore size, membranes are classified as: dense or microporous membranes ($\sim 3\,nm$), porous or macroporous membranes ($5\,nm \sim 1\,\mu m$) and fibrous membranes ($2\,\mu m$ and

Table 5.1. *Gas permeabilities through dense polymer membranes*

Membrane	Temp. (°C)	Permeability $P/10^{-10} \times \dfrac{cm^3(STP)\,cm}{cm^2 s\, cmHg}$				Diffusion coefficient $D/10^{-7} \times cm^2\,s^{-1}$				Permeability ratio		
		He	O$_2$	CO$_2$	N$_2$	He	O$_2$	CO$_2$	N$_2$	$\dfrac{P_{He}}{P_{N_2}}$	$\dfrac{P_{O_2}}{P_{N_2}}$	$\dfrac{P_{CO_2}}{P_{N_2}}$
Poly(dimethyl siloxane)	20	216	352	1120	181	600	189	189	123	1.19	1.94	6.19
Natural rubber	25	—	23.4	154	9.5	216	17.3	12.5	11.7	—	2.46	16.2
Polybutadiene	25	—	19.0	138	6.45	—	15.0	10.5	11.0	—	2.95	21.4
Ethylcellulose	25	53.4	14.7	113	4.43	22	6.4	5.65	2.33	12.0	3.31	25.6
Polyethylene (low density)	25	4.93	2.89	12.6	0.97	68	4.6	3.72	3.20	5.08	2.98	13.0
Polystyrene	20	16.7	2.01	10.0	0.315	75	22	30	3.0	53.0	6.38	31.7
Polycarbonate	25	19	1.4	8.0	0.3	—	0.21	0.048	—	633	4.7	26.7
Polyethylene (high density)	25	1.14	0.41	3.62	0.143	30.7	1.70	1.24	0.93	7.97	2.87	25.3
Poly(vinyl acetate)	20	9.32	0.225	0.676	0.032	—	—	—	—	291	7.03	21.1
Poly(vinyl chloride)	25	2.20	0.044	0.149	0.012	1.74	0.044	0.0125	0.010	191	3.83	13.0
Acethylcellulose	22	13.6	0.43	—	0.14	—	—	—	—	97.1	3.0	—
Nylon 6	30	—	0.038	0.16	0.01	—	—	—	—	—	3.8	16.0
Polyacrylonitrile	20	0.44	0.0018	0.012	0.0009	—	—	—	—	488	2.0	13.3
Poly(vinylidene chloride)	20	0.109	0.00046	0.0014	0.00012	—	—	—	—	908	3.8	11.7
Poly(vinyl alcohol)	20	0.0033	0.00052	0.00048	0.00045	—	—	—	—	7.3	1.1	1.16

above). However, the permeation behaviours of these membranes do not simply correlate with the pore size, since they are influenced also by the chemical nature of the membrane matrix, shape and distribution of pores, degree of crystallization, charged state, and so on.

Dense membranes, a category which includes ordinary polymer films, are prepared by means of direct polymerization, stretching, blast moulding, or casting. They have no distinctive pore structure, but solute molecules permeate across molecular interstices of the polymer chains or by the aid of cooperative motions of chain segments.

When the interstitial volume exceeds 40%, the interstices are combined together to form macropores. Various methods for preparation of macroporous membranes have been developed. For example, the polymeric material is dissolved in a mixture of good solvent and poor solvent and, when the good solvent is removed by vaporization, a porous gel structure is formed by phase separation. The pore structure is affected by the composition of the solvent system and the vaporization speed; by this method porous nitrocellulose membranes with pore diameters from 5 to 8000 nm can be prepared. Porous membranes having microdomain structures can be prepared of some block copolymers of A–B type or A–B–A type, since the chain segments consisting of each component gather together to form respective domains. Many membranes with microdomain structures have been prepared, an example of which is cast membrane prepared from a tetrahydrofuran solution of poly(vinylpyridine-co-styrene). Recently, the membrane materials with these multi-phase structures have found use in biomedicine, owing to their excellent biocompatibility properties.

Porous membranes may also be prepared from polyion complexes, which are formed from polyelectrolytes with different fixed charges. Their charged state is regulated by the combination of component polyelectrolytes, the mixing ratio, the media and pH; and their porosity is controlled by the charged state as well as by the membrane-forming conditions. The regular ladder structure is formed from rigid polysaccharides such as glycol chitosan hyaluronic acid, and the irregular scramble structure is formed from flexible vinyl polymers. Membranes having straight cylindrical pores have been prepared from polyelectrolyte sols by the phase-separation method, and have been named 'ionotropic-gel membranes' by Thiele. It was reported that the cupric alginate membrane thus prepared has a pore size of 1.2 nm (upper surface) and 3.5 nm (lower surface) and a pore density of 2.9×10^5 cm^{-2} (upper) and 4.0×10^4 cm^{-2} (lower).

The nucleation-track membrane is a porous membrane with a uniform and straight pore structure. When a high-energy radiation from nuclear fission products bombards a dense membrane material, straight pores are formed by chain scission owing to the resultant ionization. The linear energy loss dE/dx to form pores is 4 meV mg^{-1} cm^{-2} for a polycarbonate membrane and pores of 2.5 nm in diameter are formed in about 0.5% of the membrane surface when a membrane of $10\,\mu m$ in thickness is irradiated by the dose rate of 10^{11} neutrons cm^{-2}.

Fibrous membranes having coarse pores are formed by entanglement of fibres, and take the form of filter papers, felt, matt, unwoven cloth, and so on. According to the size or molecular weight of filtered particles, the filtration process is classified into usual filtration ($1\,\mu m$ and larger), microfiltration ($0.025 \sim 10\,\mu m$), ultrafiltration ($1 \times 10^3 \sim 3 \times 10^5$ mol. wt) and hyperfiltration (smaller molecules or ions). Reverse osmosis is classified as a kind of hyperfiltration, but the separation mechanism of this method is not simple filtration.

5.2.3 *Liquid permeation through polymer membranes*

Membrane processes of liquid permeation have been developed for practical applications such as the desalination of water. Recently, this method has been recognized as an energy-saving process, since it requires no energy for phase changes, contrary to the distillation method.

The transfer of solvent molecules through a membrane is called osmosis, whereas the process by which solvent water is transferred through a membrane from saline water by application of pressure is called reverse osmosis; in the latter case the direction of water transfer is reversed because the pressure applied exceeds the osmotic pressure.

The development of a membrane that could filter salts such as NaCl from water was already under way in the 1950s. It was found that an acetylcellulose membrane resists the transport of inorganic salts, but this membrane was not practical owing to its small water permeability. In 1960, S. Loeb & S. Sourirajan developed an asymmetric acetylcellulose membrane having a high water flux as well as a high salt rejection. Owing to this finding, the reverse osmosis method came to have a practical value as a novel seawater desalination process. This membrane is prepared by the following procedure: the casting solutions are prepared by dissolving acetylcellulose in water/acetone/Mg(ClO$_4$)$_2$ or water/formamide/acetone, and the membrane is formed by casting the solution on to a flat surface such as a glass plate followed by vaporization and immersion into an ice/water mixture. The required properties are rendered by annealing at a prescribed temperature.

The membrane thus prepared is asymmetric. The membrane consists of two layers: the dense skin layer, which has a pore size estimated to be $1 \sim 3$ nm, is $1\,\mu m$ thick or less at the upper side, while the lower layer of $80 \sim 160\,\mu m$ thickness is macroporous. Annealing causes contraction of the membrane; the higher the annealing temperature is, the lower the water flux is, and the higher the salt rejection is. The reason why the membrane rejects salts so effectively is considered to be as follows. The acetylcellulose matrix adsorbs water avidly to form a hydration layer on the surface. The structure-breaking salt ions are excluded from this layer, while water is incorporated and passes relatively easily through the pores of the membrane.

Reverse osmosis membranes with good performances are also prepared from aromatic polyamides such as:

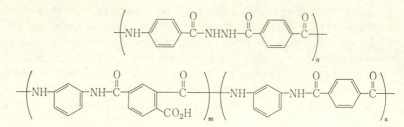

and benzimidazorone polymer:

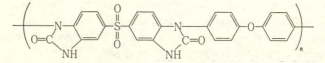

In general, the separation method driven by external pressure difference between liquid states is called the hydraulic permeation method, and it can be applied to the separation of mixtures of organic compounds. The transfer process driven by the difference in osmotic activity (or concentration) is known as diffusion and especially the diffusion process of the solvent is called dialysis. Differences in permeation rates of various compounds, arising mainly from their solubility differences in the membrane phase, are utilized as a separation method, which is called the perstraction method. It may be considered as a kind of extraction method using the membrane as an extractant.

When a membrane is placed at the interface between a vapour phase and a liquid phase during vaporization, separation takes place owing to the difference in permeation rate through the membrane. This is called a pervaporation method.

These membrane processes can be applied to the separation of a mixture containing components with nearly equal boiling points, such as an isotopic mixture, or an azeotropic mixture. Both the diffusion coefficient ratio and the solubility ratio govern the extent of separation, and the former is the primary factor in the separation of isotopic mixtures. For example, the permeation rates of *ortho-*, *meta-*, and *para*-xylene through a polyethylene membrane are 71.4, 98.3, and 125 $g \, cm \, h^{-1} \, cm^{-2}$, respectively, while the cross sections of these molecules are estimated to be 13.9, 13.2, and 12.8 $cm^2 \, mol^{-1}$, respectively.

5.2.4 *Charged membranes*

In 1950, synthetic charged membranes were prepared for the first time by W. Juda & M. R. Wyllie. Before that it had been noticed that the positively or negatively charged membranes could exhibit specific behaviour. Thus, in 1932 F. G. Donnan established the conditions of membrane equilibrium based on the Gibbs' theorem, and K. H. Meyer, J. F. Sievers & T. Teorell elucidated theoretically that a charged membrane shows permselectivity for ions according to the sign of the charge, and it was clear that electrolytic dialysis by means of charged membranes could be applied to desalination and salt concentration.

The fundamental properties of charged membranes are due to an electrostatic effect; a negatively charged membrane exhibits a cation-selective permeability and a positively charged membrane exhibits an anion-selective permeability. These membranes are commercially available as ion-exchange membranes, which are typically polystyrene membranes crosslinked with divinylbenzene having fixed sulphonate groups $-SO_3^{\ominus}$ (a cation-selective membrane) or fixed ammonium groups $-NR_3^{\oplus}$ (an anion-selective membrane). The distribution of ions inside and outside the membrane is expressed for the charged membrane of fixed charge density, $\bar{X}$, equilibrated with a 1–1 type electrolyte solution according to the Gibbs' equilibrium condition as follows,

$$\frac{\bar{f}_{\pm}^{2} x(\bar{x} + \bar{X})}{f_{\pm}^{2} c_s^{2}} = \frac{(\bar{f}_w \bar{c}_w)^{v_s/v_w}}{(f_w c_w)^{v_s/v_w}} \tag{5.8}$$

where c_s, the concentration of electrolyte solution; $f_{\pm}$, the mean ionic activity coefficient; $\bar{x}$, the concentration of coions in the membrane phase; $\bar{X}$, the fixed charge density (the bar denotes the value in the membrane phase); c_w, the molar concentration of water; f_w, the activity coefficient of water; v_s, the molar volume of the electrolyte, and v_w, the molar volume of water. This is Donnan's membrane equilibrium equation. If the properties of water in the membrane phase are the same as those of water in the bulk phase equation (5.8) reduces to

$$\bar{f}_{\pm}^{2} \bar{x}(\bar{x} + \bar{X}) = f_{\pm}^{2} c_s^{2} \tag{5.9}$$

and by solving this equation:

$$\bar{x} = \frac{-\bar{X} + \sqrt{\bar{X}^2 + 4(f_{\pm}/\bar{f}_{\pm})^2 c_s^{2}}}{2} \tag{5.10}$$

This relation shows that the coion concentration $\bar{x}$ in the membrane phase is very low when the value of $c_s/\bar{X}$ is sufficiently small. In other words, the membrane excludes ions that are of the same sign as the fixed charge, and takes up selectively ions that are of the opposite sign when the external salt concentration is sufficiently low compared to the fixed charge density. This is the origin of the ionic permselectivity of charged membranes.

Although the charged membrane shows a distinct permselectivity for ions of different signs as stated above, a membrane with a permselectivity for a given ionic species is necessary in cases where concentration or exclusion of specific ions is demanded. It is considered that the ionic permselectivity is ruled by the ionic distribution and the diffusion rate in the membrane, and the order of permselectivities of monovalent ions through the usual cation-selective membrane of the poly(styrene sulphonate) type is as follows:

$$H^{\oplus} > Cs^{\oplus} > Rb^{\oplus} > K^{\oplus} > NH_4^{\oplus} > Na^{\oplus} > Li^{\oplus}$$

The selective concentration of monovalent ions with concomitant exclusion of divalent ions are required in the salt concentration process from seawater, because the scale formation of $CaSO_4$ caused by the concentration of $Ca^{2\oplus}$ and $SO_4^{2\ominus}$ is prohibitive for stable operation. The usual ion-exchange membrane shows a higher selectivity for divalent ions compared with monovalent ions. The exclusion of polyvalent ions can be attained by depositing a thin polycation layer on to the surface of the cation-selective membrane, or a thin polyanion layer on to the surface of the anion-selective membrane. The mechanism of exclusion is considered to be due to the larger electrostatic repulsion between polyvalent ions and the same-sign charge of the polyion layer. This is a kinetically controlled mechanism.

Another method, which gives permselectivity based on the equilibrium-controlled mechanism, is to introduce the functional groups interacting with a given ionic species into the membrane. For example, it was reported that the introduction of nitro groups into a cation-selective poly(styrene sulphonate)-type membrane enhances the selectivity for $K^\oplus$ ions. However, since the larger affinity for an ionic species reduces the ionic mobility in the membrane phase, this method does not give the desired results in most cases.

Charged membranes are used as diaphragms in the various electrolysis processes. In the NaCl electrolysis process to produce caustic soda and chlorine, a diaphragm membrane that has a good performance and at the same time a long life under oxidizing and alkaline conditions at elevated temperatures near $80\,^\circ C$ is required. A membrane which fulfils these requirements is a perfluorinated ion-exchange membrane, the chemical structure of which is as follows:

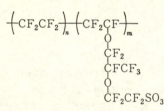

It is said that the current efficiency of caustic soda production is primarily determined by the equivalent weight, which is defined as the membrane weight per equivalent of sulphonate group. Another membrane that is produced by replacing sulphonate groups by carboxylate groups also shows good properties.

5.3   Theories of membrane transport
5.3.1   *Phenomenological equations for membrane transport processes*
Let us consider a system consisting of two phases separated by a membrane. For the moment the membrane will be considered as a two-dimensional substance having no thickness. The two phases are characterized by the intensive state variables such as temperature, T, pressure, P, electric potential, E, and chemical potential, μ_i (or concentration, c_i) of the component

i. When the values of these state properties are different across the membrane, there occur various transport phenomena such as heat conduction, electrical conduction, salt flow and volume flow as shown schematically in Fig. 5.2.

The membrane system exhibiting various transport processes is systematically analysed based on non-equilibrium thermodynamics. The dissipation function which relates to the entropy production rate in spontaneous changes is expressed for an isothermal transport process as:

$$\Phi = T\frac{d_i S}{dt} = -\sum_i J_i \Delta\mu_i \tag{5.11}$$

where $\Delta\mu_i = \mu_i^{II} - \mu_i^{I}$ is the chemical potential difference of component *i* between phase I and II and J_i is the mass flux of *i*. If we consider a simple system which consists of an electrolyte, s (cation 1 + anion 2), an unpermeable solute, i, and water, w,

$$\Phi = -(J_1\Delta\mu_1 + J_2\Delta\mu_2 + J_w\Delta\mu_w) \tag{5.12}$$

For the electrolyte,

$$\Delta\mu_s = \nu_1\Delta\mu_1 + \nu_2\Delta\mu_2, \quad J_s = J_1/\nu_1 = J_2/\nu_2 \tag{5.13}$$

and, if the electrode is reversible to the anion, the potential difference, ΔE, and the current density, I, are expressed by

$$\Delta E = \Delta\mu_2/z_2 F, \quad I = (z_1 J_1 + z_2 J_2)F \tag{5.14}$$

respectively, where z_1 is the charge number and F the Faraday constant, and then equation (5.12) reduces to

$$\Phi = -(J_w\Delta\mu_w + J_s\Delta\mu_s + I\Delta E) \tag{5.15}$$

Further, the volume flux, J_v, and the diffusion flux, J_d, are defined by

$$J_v = J_s v_s + J_w v_w \tag{5.16}$$

Fig. 5.2. Membrane phenomena.

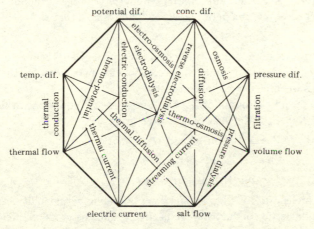

and

$$J_d = (J_s/c_s) - (J_w/c_w) \tag{5.17}$$

respectively, where v_s and v_w are the molar volume and c_s and c_w are the molar concentrations, and then

$$\Phi = -[J_v(\Delta P - \Delta \pi) + J_s \Delta \pi_s \{1 + (\phi_s \Delta \pi/\Delta \pi_s)\}/c_s + I\Delta E] \tag{5.18}$$

where $\Delta \pi = \Delta \pi_s + \Delta \pi_i$ is the osmotic pressure difference, $\Delta \pi_s$ and $\Delta \pi_i$ are the osmotic pressure differences due to the electrolyte and the unpermeable solute, respectively, and $\phi_s = c_s v_s$ is the volume fraction of the electrolyte. According to van't Hoff's equation,

$$\Delta \pi_s = RT\Delta c_s \tag{5.19}$$

and under the conditions $\phi_s \Delta \pi/\Delta \pi_s \ll 1$, equation (5.18) leads to

$$\Phi = -[J_v(\Delta P - \Delta \pi) + J_s(\Delta \pi_s/c_s) + I\Delta E] \tag{5.20}$$

These equations were derived by O. Kedem & A. Katchalsky (*Trans. Faraday Soc.* **59**, 1931, 1941 (1963)).

The dissipation function is expressed as the sum of the products of forces and fluxes which are conjugate to each other and the linear relations hold between the forces and the fluxes in a near-equilibrium range. Thus the following phenomenological equations are derived from equation (5.20)

$$\left. \begin{array}{l} J_v = -\{L_{11}(\Delta P - \Delta \pi) + L_{12}(\Delta \pi_s/c_s) + L_{13}\Delta E\} \\ J_s = -\{L_{21}(\Delta P - \Delta \pi) + L_{22}(\Delta \pi_s/c_s) + L_{23}\Delta E\} \\ I = -\{L_{31}(\Delta P - \Delta \pi) + L_{32}(\Delta \pi_s/c_s) + L_{33}\Delta E\} \end{array} \right\} \tag{5.21}$$

In general, a flux is caused not only by the conjugated force but also by non-conjugated forces, a phenomenon known as the coupling effect. The coupling effect is restricted by the symmetry requirements. The requirement of spatial symmetry (Curie–Prigogine's theorem) prohibits the coupling between a scalar process (e.g. ordinary chemical reaction) and a vector process (e.g. diffusion) and the requirement of temporal symmetry (Onsager's theorem) leads to the reciprocal relation between the coupling terms; that is,

$$L_{ij} = L_{ji} \tag{5.22}$$

Owing to the latter relation, nine phenomenological coefficients reduce to six (three intrinsic coefficients and three cross coefficients) for describing the transport process of three degrees of freedom.

For practical purposes, the following phenomenological equations, which were derived by Kedem & Katchalsky, are often utilized,

$$\left. \begin{array}{l} J_v = -L_p(\Delta P - \Delta \pi_i) + \sigma L_p \Delta \pi_s + (P_E L_p/\kappa)I \\ J_s = -c_s(1 - \sigma)J_v - \omega \Delta \pi_s - (\tau_1/\nu_1 z_1 F)I \\ I = P_E J_v - (\kappa \tau_1/\nu_1 z_1 F)(\Delta \pi_s/c_s) - \kappa \Delta E \end{array} \right\} \tag{5.23}$$

where the intrinsic coefficients L_p, ω, and κ are hydraulic permeability, solute permeability and electric conductivity, respectively; and the cross coefficients

σ, P_E and τ, are the reflection coefficient, electro-osmotic pressure and transport number, respectively. These are defined according to the following equations:

$$c_s(1-\sigma) = -\left(\frac{J_s}{J_v}\right)_{\Delta\pi_s, I} \tag{5.24}$$

$$P_E = \left(\frac{I}{J_v}\right)_{\Delta\pi_s, \Delta E} = -\left(\frac{\Delta P - \Delta\pi_i}{\Delta E}\right)_{J_v, \Delta\pi_s} \tag{5.25}$$

$$\frac{\tau_1}{\nu_1 z_1 F} = -\left(\frac{J_s}{I}\right)_{J_v, \Delta\pi_s} = \left(\frac{\Delta E}{\Delta\pi_s/c_s}\right)_{J_v, I} \tag{5.26}$$

When no current runs, the volume flux through the membrane is written from equation (5.23):

$$J_v = -L_p\{(\Delta P - \Delta\pi_i) - \sigma\Delta\pi_s\} \tag{5.27}$$

where $\Delta\pi_i$ is the osmotic pressure difference due to the unpermeable solute, which is sometimes called a colloidal osmotic pressure. As can be seen from equation (5.27), the volume flow through the membrane is primarily governed by the pressure difference and the colloidal osmotic pressure. This relation explains how water is transported through the wall of a blood vessel. Water enters into a capillary vessel at the end of a vein and flows out of the vessel at the end of an artery. The colloidal osmotic pressure of blood plasma is 36 cmH$_2$O against lymph, while the hydraulic pressure of blood is 44 cmH$_2$O at the artery's end and 17 cmH$_2$O at the vein's end. The driving force of water is 8 cmH$_2$O inwards at the artery's end and 19 cmH$_2$O outwards at the vein's end.

$\Delta\pi_s$ is the osmotic pressure due to the solute electrolyte, and this contributes to various extents to the pressure difference. When the electrolyte is completely unpermeable, the reflection coefficient, σ, is unity, and conversely, when the electrolyte is completely permeable σ is zero. For the partially permeable electrolyte, σ ranges from zero to unity; that is, the value of σ measures the extent of resistivity against solute permeation through the membrane. When a solution containing a partially permeable solute is separated from the pure solvent by the membrane, at first an osmotic pressure is set up, resulting in solvent flow according to the formula $J_v = \sigma L_p \Delta\pi_s$, and then the osmotic pressure decreases to zero owing to the leakage of the solute (see Fig. 5.3).

5.3.2 *Consideration of membrane structure*

In reality membranes have a finite thickness and an internal structure. Thus, the treatment in the preceding section is an approximation that assumes the membrane to be a 'black box'. It is necessary to analyse by taking the membrane structure into consideration in order to obtain details of the membrane properties such as a nonlinear characteristic of voltage and current.

Then, let us consider a membrane as consisting of many thin layers stacked parallel to each other. The linear phenomenological equations (5.21) are applied to each thin layer; that is,

$$\left.\begin{aligned}
\partial J_v &= -\{l_{11}\partial(P-\pi) + l_{12}(\partial\pi_s/c_s) + l_{13}\partial E\} \\
\partial J_s &= -\{l_{21}\partial(P-\pi) + l_{22}(\partial\pi_s/c_s) + l_{23}\partial E\} \\
\partial I &= -\{l_{31}\partial(P-\pi) + l_{32}(\partial\pi_s/c_s) + l_{33}\partial E\}
\end{aligned}\right\} \qquad (5.28)$$

The differential phenomenological coefficients $l_{11}, l_{12}, \ldots$ are assumed to be constants which are determined only by the membrane material, temperature, species and concentration of electrolyte. Figs. 5.4 and 5.5 show the results which are obtained by resolving the flow equations (5.28) for a membrane of thickness d, consisting of twenty ($k=20$) thin layers of equal thickness.

Fig. 5.4 shows that the linear relationship between force and flow does not always hold across the membrane, in spite of the existence of the local linear relationship. Fig. 5.5 gives the distribution of the coion concentration, the pressure and the electric potential in the membrane phase. These profiles in the membrane are not linear even when no current flows. This is because diverse forces exert themselves simultaneously and produce diverse responses. The coion concentration distribution is largely dependent on the direction of the current, and the distribution changes remarkably in the region neighbouring the concentrated solution when the current flows from the concentrated to the dilute solution ($I<0$), and does so in the region neighbouring the dilute solution when the current flows the other way ($I>0$). In the case of $I>0$, a negative pressure region is produced, because the osmotic pressure acts compensatively.

More specific properties are realized for composite membranes consisting of cation-selective and anion-selective membranes. For the composite membrane which is constructed by arranging alternately cation-selective and anion-selective membranes in series, the permeation rate of electrolyte is larger than

Fig. 5.3. Osmotic pressures shown by various solute–membrane systems. (1) Completely permeable solute ($\sigma=0$); (2) partially permeable solute ($0<\sigma<1$); (3) non-permeable solute ($\sigma=1$); (4) partially permeable and non-permeable solutes.

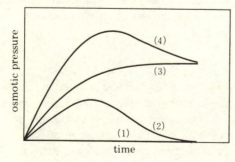

Fig. 5.4. Relationships of volume flux J_v, salt flux J_s and potential difference ΔE against current I. 0.20 M/0.02 M NaBr aqueous solution, membrane thickness 10 mm. (C. McCallum & P. Meares, *J. Membrane Sci.*, **1**, 65 (1976).)

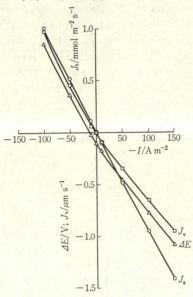

Fig. 5.5. Profiles of coion concentration (*a*), pressure (*b*) and electric potential (*c*) in membranes. 0.20 M/0.02 M NaBr aqueous solution, membrane thickness 10 mm. $I/A\,m^{-2}$: a, −150; b, −100; c, −50; d, −10; e, 0; f, 10; g, 50; h, 100, and j, 150. (C. McCallum & P. Meares, *J. Membrane Sci.*, **1**, 65 (1976).)

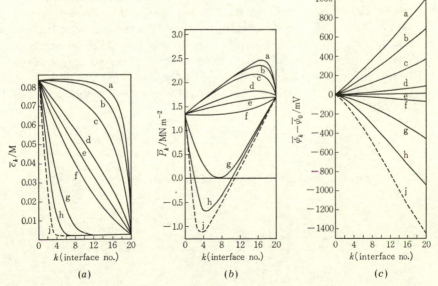

the simple sum of the rates through each portion and, therefore, it is expected that a novel mechanism of salt permeation exists. As shown schematically in Fig. 5.6, cations permeate through the cation-selective parts of the membrane and anions do so through anion-selective parts, thus transporting salt synergistically. In this case, the ionic flows through respective portions form a circulating current. This mechanism of salt permeation is utilized as pressure dialysis through a mosaic composite membrane. On the other hand, for a composite membrane that is constructed by stacking alternately cation-selective and anion-selective membranes in parallel, salt accumulation and depletion occur at the interface between different portions and, therefore, a remarkable polarizing effect is observed in the ionic transport and the electric conduction behaviour. For example, a distinct electric rectification effect is displayed by a parallel composite membrane made up of an $H^{\oplus}$-form cation-selective membrane and an $OH^{\ominus}$-form anion-selective membrane. Various composite transport functions would be expected to occur in biological membranes since they have diverse composite structures.

5.3.3 *Mechanisms of membrane transport*

Transport behaviour through membranes is classified into simple transport, facilitated transport, active transport, and so on, according to the type of mechanism involved. The examples described in the preceding sections are regarded as cases of simple transport. Transport through a membrane having no binding site is caused by the solution–diffusion mechanism, and transport through a membrane having fixed binding sites occurs by the jump mechanism. Mobile binding sites serve as carriers and the transport by these carriers is called facilitated transport.

A typical example is the transport of oxygen by haemoglobin in blood. P. E. Scholander measured the transport rates of O_2 and N_2 through an aqueous haemoglobin solution by changing the pressures at both the sides of the aqueous membrane phase (liquid membrane), while the pressure difference was fixed. A result is shown in Fig. 5.7, which shows clearly the occurrence of carrier transport of oxygen. By representing O_2, Hb (haemoglobin) and HbO_{2n} (haemoglobin-O_2 adduct) by 1, 2 and 3, respectively, the transport processes are given by

$$J_1 = -L_1 \Delta\mu_1, \quad J_2 = -L_2 \Delta\mu_2, \quad J_3 = -L_3 \Delta\mu_3 \tag{5.29}$$

Fig. 5.6. Pressure dialysis through a mosaic composite membrane.

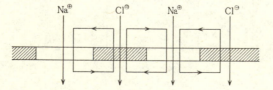

where J_i, μ_i and L_i are the flux, the chemical potential and the phenomeno-logical coefficient of species i, respectively. The dissipation function is expressed by

$$\Phi = -J_1\Delta\mu_1 - J_2\Delta\mu_2 - J_3\Delta\mu_3 \tag{5.30}$$

By using equilibrium conditions,

$$A = n\mu_1 + \mu_2 - \mu_3 = 0, \quad K = c_1{}^n c_2/c_3 \tag{5.31}$$

for the chemical reaction $nO_2 + Hb = HbO_{2n}$, equation (5.30) reduces to:

$$\Phi = -(J_1 + nJ_3)\Delta\mu_1 - (J_2 + J_3)\Delta\mu_2 \tag{5.32}$$

from which the following phenomenological equations are derived:

$$J_1{}^* = J_1 + nJ_3 = -L_{11}\Delta\mu_1 - L_{12}\Delta\mu_2$$
$$= -(L_1 + nL_3)\Delta\mu_1 - nL_3\Delta\mu_2 \tag{5.33a}$$
$$J_2{}^* = J_2 + J_3 = -L_{21}\Delta\mu_1 - L_{22}\Delta\mu_2$$
$$= -nL_3\Delta\mu_1 - (L_2 + L_3)\Delta\mu_2 \tag{5.33b}$$

Since there is no net flow of haemoglobin,

$$J_2{}^* = -nc_3 u_3 RT\Delta \ln c_1 - (c_2 u_2 + c_3 u_3)RT\Delta \ln c_2$$

$$= -u_3 RT\left\{\frac{nc_1{}^{n-1}c_2}{K}\Delta c_1 + \left(1 + \frac{c_1{}^n}{K}\right)\Delta c_2\right\} = 0 \tag{5.34}$$

In this derivation, the relations $\Delta\mu_i = RT\Delta \ln c_i$ and $L_i = c_i u_i$ are used, where u_i is the mobility and u_2 is approximately equal to u_3. Equation (5.34) leads to,

$$\Delta \ln c_2(K + c_1{}^n) = \Delta \ln K(c_2 + c_3) = 0 \tag{5.34'}$$

This relation means that the concentrations of Hb and HbO_{2n} change from place to place, but the sum remains constant.

Fig. 5.7. Transport of oxygen through a haemoglobin solution. (P. F. Scholander, *Science*, **131**, 585, 1379 (1960).)

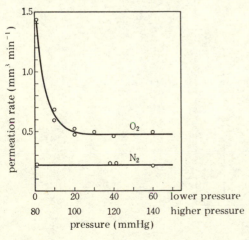

The total flow of oxygen is expressed by:

$$J_1^* = -u_1 RT \Delta c_1 - n u_2 RT \left(\frac{c_1^{n-1} c_2}{K} \Delta c_1 + \frac{c_1^n}{K} \Delta c_2 \right)$$

$$= -u_1 RT \Delta c_1 + n u_2 RT \Delta c_2 \tag{5.35}$$

By using the diffusion coefficient $D_i = u_i RT$, and equation (5.34'),

$$J_1^* = -D_1 \Delta c_1 + n D_2 \Delta c_2$$

$$= -D_1 \Delta c_1 - n D_3 \Delta c_3 \tag{5.36}$$

The second term of the last equation represents the flow of oxygen facilitated by haemoglobin, which is saturated by an increase in oxygen pressure even when the pressure difference is exerted across the liquid membrane. When the oxygen pressure is low, carrier transport predominates simple transport, and so oxygen is selectively transported.

Macrocyclic polyethers, often called crown ethers, are utilized as carriers of metallic ions and have served as model compounds of ionophores, which are transport carriers of metallic ions in living systems. The macrocyclic polyether has a molecular cavity surrounded by ether oxygens, which will combine strongly with a metallic ion of a suitable ion radius. The permeation rates of various cations across a liquid membrane of chloroform solution of dibenzo-18-crown-6 (2,3,11,12-dibenzo-1,4,7,10,13,16-hexaoxacyclooctadeca-2,11-diene) are shown in Fig. 5.8. Monovalent cations ($K^\oplus$, $Ag^\oplus$) and bivalent cations ($Pb^{2\oplus}$, $Hg^{2\oplus}$) are selectively transported across this liquid membrane. The ionic radii of these cations are 0.12~0.13 nm, and fit well into the cavity radius of the crown ether.

Fig. 5.8. Selective transport of metal ions through a liquid membrane containing dibenzo-18-crown-6. (C. F. Reusch & E. L. Cussler, *AIChE J.*, 19, 736 (1973).)

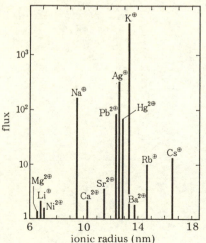

This facilitated transport is caused by the carrier crown ether, C. If the carrier complex C-K$^\oplus$ formed by

$$K^\oplus + C \rightleftharpoons C\text{-}K^\oplus$$

migrates through the liquid membrane, the total flow of K$^\oplus$ across the membrane is expressed by

$$J_K* = \frac{D_K K_d}{l} \Delta c_K + \frac{D_{cK} K_d Kc c_K}{l} \Delta \left(\frac{c_K}{1 + K_d Kc_K} \right) \tag{5.37}$$

where K_d is the distribution coefficient, K is the stability constant of the complex and D is the diffusion coefficient. But the observed result is not consistent with equation (5.37). The reason is that the carrier complex C-K$^\oplus$ is accompanied by a counter ion Cl$^\ominus$ in the membrane; that is, the formation of the carrier complex by

$$K^\oplus + Cl^\ominus + C \rightleftharpoons C\text{-}K^\oplus \cdot Cl^\ominus$$

is involved in the transport process. In this case, the total flow of K$^\oplus$ is expressed by

$$J_K* = \frac{D_K K_d}{l} (\Delta c_K)^2 + \frac{D_{cK} K_d Kc c_K}{l} \Delta \left(\frac{c_K^2}{1 + K_d Kc_K^2} \right) \tag{5.38}$$

This equation is consistent with the observed result.

Under conditions where the simple transport given by the first term of the right-hand side of equation (5.38) is negligibly small, the transport is caused predominantly by the carrier, being governed by the association strength between the carrier and ions, and then a highly selective transport system can be realized. For example, it was reported that the enantiomer separation of the optically active ammonium salt can be achieved by means of a crown ether of RR con-

$$p\text{-OHC}_6H_4CHCO_2CH_3$$
$$\overset{|\oplus}{Cl^\ominus NH_3}$$

figuration as carrier, (M. Newcomb, R. C. Helgeson & D. J. Cram, *J. Amer. Chem. Soc.*, **96**, 7367 (1974)).

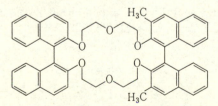

We can construct a transport system in which a solute is transported against its concentration gradient. Two kinds of solutes A and B are contained in the solutions I and II, which are separated from each other by a liquid membrane containing a carrier substance, C. The carrier combines with both A and B; let

$c_A{}^I < c_A{}^{II}$, and $c_B{}^I \ll c_B{}^{II}$. Owing to a coupling effect caused by the existence of a common carrier, an intense flow of B from II to I can drive an uphill flow of A from I to II. This is a pumping process of A by B. The total flux of A is represented by

$$J_A{}^* = J_A + J_{AC}$$

$$= \frac{D_A K_{dA}}{l} \Delta c_A + \frac{D_{AC} K_{dA}}{l} (R + K_{dB} K_B \bar{c}_B) \Delta c_A$$

$$- \frac{D_{BC} K_{dA}}{l} R K_{dB} K_B \bar{c}_A \Delta \bar{c}_B \tag{5.39}$$

where $\bar{c}_i = (c_i{}^I + c_i{}^{II})/2$ is a mean concentration. In the latter expression of equation (5.39), the first term is the simple transport, the second term is the ordinary carrier transport and the last term is the pumping transport.

A transport system driven by such a mechanism is realized by separating a 0.1 M NaOH solution (I) from 0.1 M NaCl plus HCl mixed solution (II) using an octanol solution of monensin. (E. M. Choy, D. F. Evans & E. L. Cussler, *J. Am. Chem. Soc.*, **96**, 7085 (1974).) The up-hill transport of $Na^\oplus$ from I to II is driven by the down-hill transport of $H^\oplus$ from II to I. This transport system exhibits a selectivity which is attributed to monensin. Thus, the selectivity of $Na^\oplus$ to $K^\oplus$ is 3:1 in the case of a 0.1 M NaCl and KCl solution (I) and a 0.1 M HCl solution (II), and the selectivity of $Na^\oplus$ to $Cs^\oplus$ is 4:1 under similar conditions.

While down-hill transport is an energy-releasing process, up-hill transport is an energy-consuming one, and then the energy of transport must be supplied by another energy source. Such a process is called an active transport, in contrast to a passive transport which is energy-releasing. Active transport is driven by coupling with another transport process (osmo-osmotic coupling) or chemical reactions (chemi-osmotic coupling). The latter is called the active transport in a narrow sense. The transport of amino acids by an ammonium salt as shown in Fig. 5.9 serves as an exemplary case. When a 0.1 M KOH

Fig. 5.9. Transport of amino acid through a liquid membrane containing ammonium salt. $N^\oplus$ denotes tricaprylmethylammonium ion.

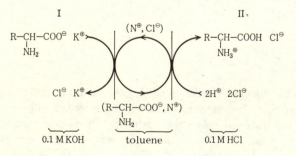

solution of amino acid (I) and a 0.1 M HCl solution (II) are separated by a liquid membrane of a toluene solution containing tricaprylmethyl-ammonium chloride, the amino acid is transported from I to II. (J. P. Behr & J.-M. Lehn, *J. Amer. Chem. Soc.*, **95**, 6108 (1973).) In the liquid membrane the exchange transport of amino acid ions and chloride ions occurs under the action of ammonium cations as a carrier and the neutralization reaction takes place at the interface of the membrane to the solution II, while the exchange reaction occurs at the interface to the solution I.

It was reported that a lecithin membrane including haematoporphyrin Mn(III) (I. Tabushi & M. Funakura, *J. Amer. Chem. Soc.*, **98**, 4684 (1976)) and liquid membranes containing vitamin K_1, ubiquinone, plastoquinone or coenzyme Q_{10} (S. S. Anderson, L. G. Lyle & R. Paterson, *Nature, Lond.* **259**, 147 (1976)) act as electron-transport membranes. The electron-transport membrane, driven by light, is constructed as shown in Fig. 5.10. (J. J. Grimaldi, S. Boileau & J.-M. Lehn, *Nature, Lond.* **265**, 229 (1977).) The nature of the electron carrier determines what is transported by the coupling with electron transport. For example, proton transport is coupled with the electron transport by a quinone carrier and anion transport is coupled with electron transport by a ferrocene-type carrier, since its reduced form is neutral and its oxidized form is positively charged. On the other hand, cation transport is coupled with the electron transport by a carrier whose reduced form is negatively charged and its oxidized

Fig. 5.10. Electron-transport membrane coupled with light. Reduced species $MV^{\ominus}$ in the reduced phase is produced from methylviologen $MV^{2\oplus}$ by proflavin (PF)-sensitized photo-reduction of EDTA (an electron donor). Electron transport is caused by oxidation of ferricyanide coupled with transport of vitamin K_3 (Q) in the liquid membrane.

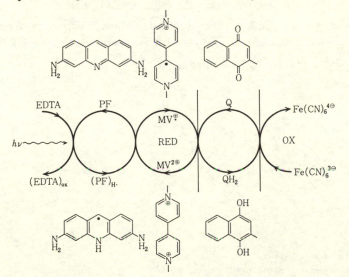

$$H_5C_6 \diagdown_{C} \diagup^{S} \diagdown / ^{S} \diagdown_{C} \diagup^{C_6H_5}$$

Ni

$$H_5C_6 \diagup^{C} \diagdown_{S} / \diagdown^{S} \diagup^{C} \diagdown_{C_6H_5}$$

form is neutral, such as nickel bisdithiolene. (J. M. Grimaldi & J.-M. Lehn, *J. Amer. Chem. Soc.*, **101**, 1333 (1979).) The rate of this cation transport process driven by the oxidation–reduction reaction is regulated by the addition of a cation carrier; for example, the addition of biscyclohexyl-18-crown-6 accelerates $K^{\oplus}$ transport and retards $Na^{\oplus}$ transport.

5.4 High molecular weight substances with transport functions

5.4.1 *Ion carriers*

Some membrane-active substances are known to serve as ion-carriers through biological membranes. They transport alkali and alkaline earth ions in the form of complexes through lipid membranes and have been named ionophores by B. C. Pressman.

(i) *Cyclic depsipeptides.* The chemical structure of valinomycin is *cyclo*(D-Val-L-Lac-L-Val-D-Hyiv)$_3$, where Val, Lac and Hyiv denote the residues of valine, lactic acid and α-oxyisovaleric acid, respectively. In solution it assumes a conformation having a central cavity of which the diameter is 0.6~0.7 nm, as shown in Fig. 5.11. This cavity size is a good fit for a $K^{\oplus}$ ion and the affinity of valinomycin for $K^{\oplus}$ is 10^4 times larger than that for $Na^{\oplus}$. The steric conformation changes on complex formation, by which $K^{\oplus}$ ion combines to six carbonyl oxygens of valine residues to strengthen the ring stability. Enniatin is *cyclo*(L-MeVal-D-Hyiv)$_3$, where MeVal denotes a methylvaline residue. The affinity to $K^{\oplus}$ ion is only 10 times larger than that to $Na^{\oplus}$ ion, because the complex structure is more liable than the valinomycin complex.

(ii) *Peptides.* Antamanide is a cyclic decapeptide which is isolated from green mushroom (*Amanita phalloides*). Its chemical structure is *cyclo*(Pro-Ala-Phe-

Fig. 5.11. Structure of valinomycin.

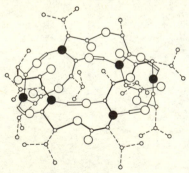

⇐ hydrogen bond
o carbon, ● nitrogen, ○ oxygen

Pro-Pro-Phe-Val-Pro). It combines with alkali metal ions to form an oil-soluble complex and shows a specificity for $Na^{\oplus}$ ion. Alamethicin is a cyclic octadecapeptide having many methylalanine residues and also forms oil-soluble complexes with alkali metal ions.

Gramicidin A is a linear peptide and its chemical structure is:

HCO-L-Val-Gly-L-Ala-D-Leu-L-Ala-D-Val-L-Trp-D-Leu-

L-Trp-D-Leu-L-Trp-NHCH$_2$CH$_2$OH.

Gramicidins B and C are homologues in which the eleventh amino acid residue is phenylalanine and tyrosine, respectively. These compounds are not carriers but channel-forming substances.

(iii) *Macrotetrolides*. These are optically active cyclic ester-ethers and frequently called actines, in which the following compounds are involved:

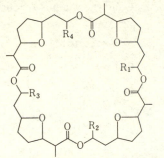

$R_1, R_2, R_3, R_4 = CH_3$ nonactin
$R_1, R_2, R_3, R_4 = C_2H_5$ tetranactin
$R_1, R_2, R_3 = CH_3, R_4 = C_2H_5$ manactin
$R_1, R_3 = CH_3, R_2, R_4 = C_2H_5$ dinactin
$R_1 = CH_3, R_2, R_3, R_4 = C_2H_5$ trinactin

The polar cavities of these compounds take up selectively $K^{\oplus}$ ions.

(iv) *Polyether carboxylic acids*. The ionophores of this type, some of which are known as antibiotics such as nigericin and monensin, have a common structure of a linear chain of five- and/or six-membered cyclic ethers having carboxylic and hydroxyl groups at both the ends of the chain. The selectiviey of monensin is $Na^{\oplus} \gg K^{\oplus} > Rb^{\oplus} > Li^{\oplus} > Cs^{\oplus}$, while the selectivity of nigericin is $K^{\oplus} > Rb^{\oplus} > Na^{\oplus} > Cs^{\oplus} > Li^{\oplus}$. Salinomycin has also a strong affinity for $K^{\oplus}$ and $Rb^{\oplus}$.

As regards ionophores effective for divalent cations, X537A and A23187 are well known and have the following chemical structures:

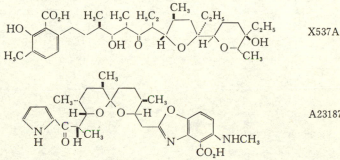

The selectivity of X537A is $Ba^{2\oplus} \gg Sr^{2\oplus} > Ca^{2\oplus} > Mg^{2\oplus}$ and the selectivity of A23187 is $Ca^{2\oplus} > Mg^{2\oplus} > Sr^{2\oplus} > Ba^{2\oplus}$. The former is effective also for alkali metal ions, but the latter is not effective for $K^{\oplus}$ ion. Free X537A exists as a dimer, which forms a 2:1 complex with $Ba^{2\oplus}$ ion. The $Ca^{2\oplus}$ ion is surrounded by seven ligands: carboxylic oxygens, carbonyl oxygens, benzoxazole nitrogens of both the molecules, and a water molecule.

Many attempts have been made to prepare model substances with functions similar to those of natural ionophores. For this purpose various macrocyclic polyethers (crown compounds) and cyclic polypeptides have been tested as candidates for ionophore-related substances. As an example, the carrier properties of model substances for A23187 are shown in Fig. 5.12.

It is said that the transport of materials through biological membranes is caused by carriers or through channels. Typical examples of the latter are the channel formed by gramicidin through a lipid membrane and the channel through an

Fig. 5.12. Facilitated transport of $Ca^{2\oplus}$ ion through a chloroform–10% ethanol liquid membrane. (W. Wierenga, B. R. Evans & J. A. Woltersom, *J. Amer. Chem. Soc.*, **101**, 1334 (1979).)

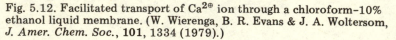

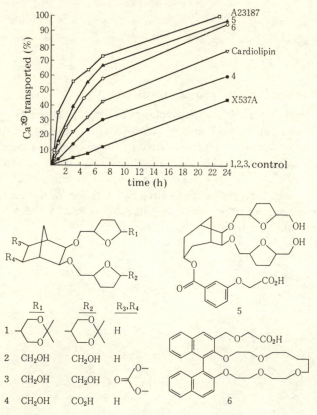

erythrocyte membrane caused by a protein known as Band 3 (because of its migration position during electrophoretic separation).

These transport functions which are specific to biological membranes are analysed on the basis of the carrier-mediated transport mechanism, as shown schematically in Fig. 5.13. The transport process consists of the following steps: (1) The carrier, C, combines with the substrate, S, outside the membrane. The extent of combination is governed by the dissociation constant at this point (K_d^{entry}) and this step is energy-releasing. (2) The association complex, CS, migrates through the membrane. In many cases, this step is driven by other energy sources. (3) Inside the membrane, the complex CS converts to C′S. The dissociation constant (K_d^{exit}) of this complex differs from K_d^{entry} and C′S dissociates to C′ and S. (4) The carrier C′ translocates to outside of the membrane and changes again into C. This transport process involves the following essential elementary processes: (i) the formation of CS complex with the function of substrate recognition, (ii) translocation of CS complex in membrane, (iii) the dissociation of CS complex caused by structural changes, and (iv) the coupling with energy-supplying processes.

The complex-formation process obeys the Michaelis–Menten equation; the transport rate of the substrate is represented by

$$v = V_{max} \left(\frac{[S_{out}]}{[S_{out}] + K_d^{entry}} - \frac{[S_{in}]}{[S_{in}] + K_d^{exit}} \right) \tag{5.40}$$

If the process is not coupled with an energy-supplying process and the relation $K_d^{entry} = K_d^{exit}$ holds, equation (5.40) reduces to

$$v = \frac{V_{max}([S_{out}] - [S_{in}])K_d}{([S_{out}] + K_d)([S_{in}] + K_d)} \tag{5.41}$$

The transport rate of the substrate is proportional to the concentration difference: it is a simple carrier transport.

If the transport process is coupled energetically, the following relationship holds under steady conditions,

Fig. 5.13. Model of carrier-mediated transport.

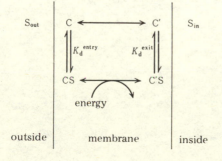

$$\frac{[S_{in}]}{[S_{out}]} = \frac{K_d^{exit}}{K_d^{entry}} \tag{5.42}$$

In other words, the concentration difference is maintained by the difference in affinity of the carrier across the membrane. Concentration differences up to $100 \sim 10\,000$ times are actually observed in many microorganism cells, and this shows the influence of a mechanism of unsymmetrical energy coupling.

5.4.2 Biological membrane-active substances

In biological systems, rather complicated but highly ordered substances play an important role in transport processes. A few examples should be noted, because they are useful for designing synthetic transport systems hereafter. In a preceding section, the transport carriers for biologically important ions were described. Carriers for amino acids and saccharides have also been isolated from microorganisms. Most of them are simple proteins having molecular weights of about 35 000 and they combine reversibly with their respective substrates. Up till now, proteins with selective transport functions for each of the following substances are known and called binding proteins: monosaccharides such as galactose, glucose, arabinose and glucose-1-phosphate; amino acids such as leucine, histidine, arginine, lysine, cysteine, glutamine, glutamic acid, proline and tryptophan, and thiamine.

Little is known about how these binding proteins effect transport processes. A model for the galactoside transport system is depicted in Fig. 5.14, but the molecular mechanism involved is not clear. The binding protein plays an essential part in the transport system, and it is required that each binding protein is selectively absorbed and fixed on to the specific membrane site; and the binding protein, after releasing the substrate, is recovered and translocated under the action of a translocating factor. This process is schematically represented in Fig. 5.15.

Fig. 5.14. Proposed galactoside transport system. Galactoside (gal-OR) is transported by a combining protein M, which changes into M* coupled with ATP → ADP to release galactoside.

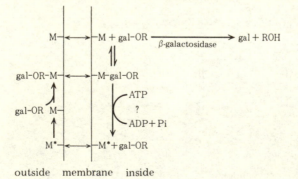

S. Roseman found a process by which saccharide is taken into the cell after phosphorylation. This is called a group translocation, an example of which, a phosphotransferase model, is schematically shown in Fig. 5.16. This process is the transport of saccharide after phosphorylation by phosphotransferase followed by the consumption of phosphoenolpyruvic acid as an energy source. It is believed that *Escherichia Coli* takes up glucose, fructose, mannose, hexosamine and β-glucoside by this mechanism.

The sodium pump, which is an active transport system for $Na^{\oplus}$ ions by the action of $Na^{\oplus}$, $K^{\oplus}$-ATPase, plays a very important role in many functions of living systems. The action of the pump is schematically shown in Fig. 5.17. ATPase is an enzyme for hydrolysis of adenosine triphosphate (ATP), and J. C. Skou found that it exists widely in plasma membranes. $Na^{\oplus}$ ions are transported against a concentration gradient to maintain the concentration difference across the membrane at a given level.

For the erythrocyte, the $Na^{\oplus}$ concentration inside and outside the membrane, $[Na^{\oplus}]_{in}$ and $[Na^{\oplus}]_{out}$, are 10 mM and 145 mM, respectively, while $[K^{\oplus}]_{in}$ and $[K^{\oplus}]_{out}$ are 150 mM and 5 mM, respectively. Then, the work required for the efflux of 3 mol $Na^{\oplus}$ is $3RT \ln (145/10) = 20.63$ kJ, and the work required for

Fig. 5.15. A model of combining protein. S, substrate; B, combining protein; M, combining site; T, translocation factor.

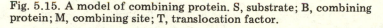
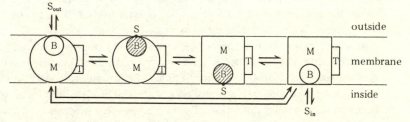

Fig. 5.16. Phosphotransferase model (a molecule translocation system). Phosphoenolpyruvic acid (PEP) phosphorylates enzyme I, which then phosphorylates protein HPr. Enzyme I and protein are non-specific to the substrate saccharide S. The phosphorylated protein HPr-P takes in saccharide S by phosphorylation catalysed by enzyme II.

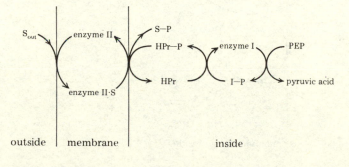

the influx of $2\,mol\,K^{\oplus}$ is $2RT\ln(150/5) = 17.49\,kJ$. The energy necessary for this work is supplied with a high efficiency by the free energy change due to hydrolysis of ATP, which is $54.39\,kJ\,mol^{-1}$ under normal conditions, that is $[ATP] = 1.5\,mM$, $[ADP] = 0.32\,mM$, $[P_i] = 0.36\,mM$. This pump operates reversibly; it was confirmed that ATP is formed from ADP and Pi (inorganic phosphoric acids) under extreme conditions such as $[Na^{\oplus}]_{out}/[Na^{\oplus}]_{in} = 150$ and $[K^{\oplus}]_{in}/[K^{\oplus}]_{out} = 510$. The action of this pump is damaged by addition of ouabain.

The absorption of saccharides and amino acids in the intestinal canal is also an active transport process against a concentration gradient, and is largely dependent on the $Na^{\oplus}$ concentration outside the membrane. The following absorption mechanism is proposed: saccharides and amino acids are transported, coupled with the $Na^{\oplus}$ flow, while the $Na^{\oplus}$ concentration inside the membrane is maintained at a low level by the action of Na,K-ATPase. This transport mechanism is a kind of cotransport.

$Ca^{2\oplus}$ ions participate in muscular contraction. The sacroplasmic reticulum in muscle cells takes up $Ca^{2\oplus}$ ions by coupling to ATP hydrolysis and then the muscle is in a contracted state. On excitation, $Ca^{2\oplus}$ ions are released and the muscle relaxes. The contraction mechanism is triggered by the binding of $Ca^{2\oplus}$ to troponin.

ATP is a very important substance which participates essentially in various energy conversion processes in living systems. On the other hand, it is widely believed that a membrane transport process is extensively related to the formation process of ATP. Thus, the chemi-osmotic coupling theory of ATP synthesis was proposed by P. Mitchell and has been supported by many observed results. In mitochondria, the organelles which carry out respiration in eucaryotes, ATP is synthesized by oxidative phosphorylation. On respiration, NADH (nicotinamide adenine dinucleotide) transports $H^{\oplus}$ to the inside of the inner mitochondrial membrane by the action of oxidation-reduction enzymes in the

Fig. 5.17. Sodium pump of $Na^{\oplus}$, $K^{\oplus}$-ATPase.

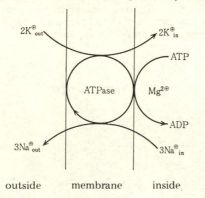

outside　　membrane　　inside

membrane; in this process, each of three traverse steps of a pair of electrons transports two $H^{\oplus}$ ions. By the resulting $H^{\oplus}$ concentration difference and electric potential difference, $H^{\oplus}$ ions pass through the ATPase F_1-F_0 complex in the inner membrane, where ATP is synthesized. 1 mol of ATP is prepared by coupling with the osmotic transport of 2 moles of $H^{\oplus}$ ions.

A similar process is involved in the photosynthesis process. ATP is synthesized by photophosphorylation in choroplast. Water is split by light and each of two traverse steps of a pair of electrons transports two $H^{\oplus}$ ions inwards. ATP is synthesized by the coupling with $H^{\oplus}$ transport through the CF_1-F_0 complex owing to the concentration and potential differences. These ATP synthesis processes are shown schematically in Fig. 5.18. The synthetic process coupled to the osmotic process is very interesting owing to its high efficiency and selectivity.

5.4.3 *Highly functional substances for transport*

Haemoglobin, an oxygen carrier in living systems, is a typical highly functional substance for transport, of which the structure and properties have been investigated in detail. Haemoglobin A in the adult erythrocyte consists of four polypeptide units: two α chains constructed of 141 amino acid residues, and two β chains constructed of 146 amino-acid residues. α and β chains form folded structures similar to each other, and gather together to form a haemoglobin molecule as shown in Fig. 5.19. Haem combines to each unit in hydrophobic pockets near the surface.

The function of haemoglobin is to transport O_2 and CO_2 by the exchange transport mechanism, and its oxygen adsorption characteristics are shown in Fig. 5.20 along with the adsorption isotherm of myglobin. Myoglobin is a haem protein in muscles and has a structure similar to that of the β chain unit of haemoglobin. Haemoglobin transports oxygen in blood vessels and myoglobin transports oxygen from capillary vessels to tissue; that is, it combines

Fig. 5.18. Model of ATP synthesis according to chemi-osmotic hypothesis: (*a*) oxidative phosphorylation; (*b*) photophosphorylation.

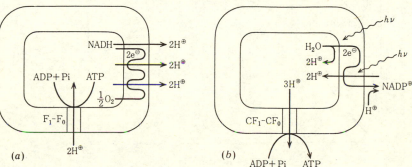

with oxygen released from haemoglobin at the ends of capillary vessels and transports it to mitochondria.

The high functionality of haemoglobin, which is correlated to that of myoglobin, is based on the characteristic S-shaped adsorption isotherm as shown in Fig. 5.20. This is caused by the interhaem interaction, which is considered as a cooperative, or allosteric, effect.

The molecular mechanism of the interhaem interaction has been worked out in detail by M. F. Perutz. Haemoglobin can assume two different stable structures, one of which is a relaxed structure, R, with strong affinity to oxygen and the other is a tense structure, T, with a limited affinity to oxygen. The T structure changes gradually to the R structure on binding with oxygen. Typically, deoxyhaemoglobin assumes the T structure and oxyhaemoglobin takes on the

Fig. 5.19. Structure of haemoglobin.

Fig. 5.20. Adsorption isotherms of oxygen on to haemoglobin and myoglobin.

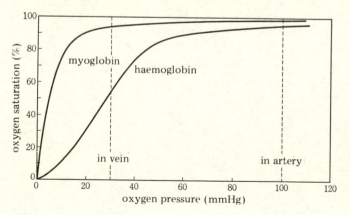

R structure; the change from the T structure to the R structure is triggered by the oxygen adsorption, as shown in Fig. 5.21.

Efforts to obtain synthetic substances to reproduce the properties of haem proteins have been pursued intensively. Since haem itself is oxidized irreversibly, the first aim is to prepare a substance that adsorbs and desorbs oxygen reversibly. The irreversible oxidation of haem is due to the charge separation of adsorbed oxygen followed by dimerization of haem crosslinked by oxygen and, therefore, it is considered that a method to prevent haem from irreversible changes is to make the adsorption site too hydrophobic for reaction with each other, or to prevent two haems from approaching each other. Along these lines, the picket fence porphyrin, the capped porphyrin and the porphyrin with side-chain ligands have been synthesized by chemical modifications on the porphyrin ring, and the haems coordinated to poly(4-vinylpyridine) and poly-L-lysine have been prepared. It was confirmed that these substances adsorb and desorb oxygen reversibly, but their functions are very simple compared to those of haemoglobin. Further efforts should be continued to reproduce the highly functional substances in Nature.

Fig. 5.21. Trigger mechanism from the T structure to R structure of haemoglobin. The adsorption of oxygen on to the haem of the T structure (thick lines) pulls the iron atom down by 0.06 nm, which interacts with the histidine residue nearby. This triggers the change to the R structure (thin lines). (M. F. Perutz, *Science (Japan)*, 9, Feb., 16 (1979).)

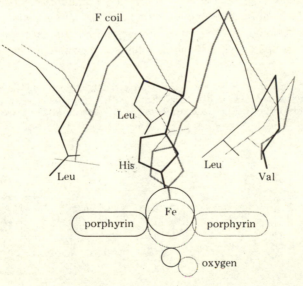

6

SYNTHETIC BILAYER MEMBRANES

TOYOKI KUNITAKE

6.1 Discovery of the 'Synthetic bilayer membrane'

The biomembrane is the most fundamental organizational unit of the living system and is directly involved in most of the basic biological functions of cells such as compartmentalization, energy transduction and information transfer. The major structural components of the biomembrane are lipid bilayers and proteins, and the fluid mosaic model of Singer & Nicolson is considered to explain its characteristic properties. In this model, protein molecules are wholly or partially immersed in the hydrophobic core of the lipid bilayer. Some of the proteins (peripheral proteins) are positioned at the hydrophilic membrane surface. Fig. 6.1 illustrates the various modes of protein binding to the lipid bilayer.

In 1965, A. Bangham and coworkers of Cambridge extracted phospholipids from egg yolk, and prepared their aqueous solution by sonication. When they examined this solution by electron microscopy, they found that bilayer vesicles were formed. This finding clearly indicated that the lipid bilayer defines the structural (organizational) characteristics of the biomembrane. Liposomes are obtainable from natural or synthetic phospholipids and possess an inner water

Fig. 6.1. Diagram illustrating the principal arrangement of the lipid bilayer and proteins in biomembranes. A: membrane-bound protein, B: amphiphilic protein, B': amphiphilic glycoprotein, C: hydrophobic protein.

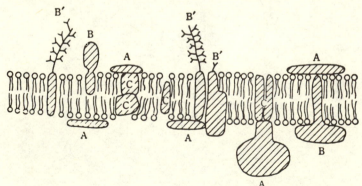

core with a diameter ranging from 30 nm (single-walled liposome) to 200 nm (multi-walled liposome), as schematically shown in Fig. 6.2. The chemical structure of these phospholipids is shown below.

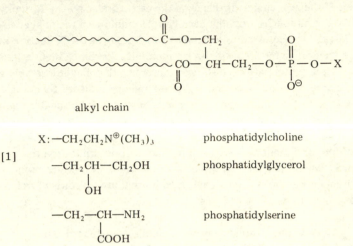

alkyl chain

[1]

X:	$-CH_2CH_2N^{\oplus}(CH_3)_3$	phosphatidylcholine
	$-CH_2CH-CH_2OH$ $\qquad \vert$ $\qquad OH$	phosphatidylglycerol
	$-CH_2-CH-NH_2$ $\qquad\quad \vert$ $\qquad\quad COOH$	phosphatidylserine

The amphiphilic nature of the phospholipid is derived from the presence of two, higher fatty acid units from C_{12} to C_{18} (more or less unsaturated) as the hydrophobic moiety, and phosphate derivatives as the hydrophilic moiety.

The phospholipid bilayer possesses many peculiar physicochemical characteristics which are essential for its biochemical functions. In particular, the acyl chain is highly oriented and can be in a crystalline or liquid-crystalline state. The reversible transition between these two phases leads to a change in the membrane fluidity, an important factor in the control of biological function.

The structural factors of phospholipids that are crucial to bilayer formation have been discussed by several authors who have suggested that the particular

Fig. 6.2. Liposomes: (*a*) multi-walled liposome, (*b*) single-walled liposome.

(*a*)
(*b*)

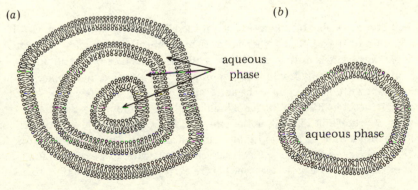

aqueous phase

aqueous phase

structure of phospholipids is responsible for formation of stable bilayers. On the other hand, we postulated that bilayer formation might be attributed to the double-chain nature of phospholipid amphiphiles, which restricts the molecular motion of the acyl chains. If this were the case, totally synthetic dialkyl compounds with simple hydrophilic head groups should give the bilayer assembly. This postulate was proved to be true by our finding of the formation of the bilayer membrane by simple dialkylammonium salts. Since then, a variety of dialkyl amphiphiles have been shown to produce the bilayer assembly in water. Furthermore, the double-chain alkyl group may be replaced by a single-chain unit with a rigid segment; and a monolayer assembly is obtainable from two-headed, single-chain amphiphiles. Thus, it has been definitely shown that membrane (bilayer or monolayer) formation is a general physicochemical phenomenon observable for certain classes of amphiphile, without limitation to specific biolipids. The term 'synthetic molecular membrane' is used for monolayer and bilayer membranes of synthetic amphiphiles that are not directly related to membrane-forming phospholipids.

6.2   Membrane-forming compounds and aggregate morphology

Synthetic amphiphiles capable of forming molecular membranes number more than 300, as of August, 1981. This is a commendable figure, if one recalls that the first example appeared only in 1977. The relation between the chemical structure of an amphiphile and its aggregate morphology will now be explained.

6.2.1   *Bilayer formation by dialkyl amphiphiles*

Double-chain ammonium salts [2] belong to a group of typical, bilayer-forming amphiphiles.

$$CH_3(CH_2)_{n-1} \diagdown \overset{\oplus}{N} \diagup CH_3$$
$$CH_3(CH_2)_{n-1} \diagup \diagdown CH_3 \qquad Br^{\ominus}$$

$$2C_n N^{\oplus} 2C_1 \qquad [2]$$

They produce stable bilayer assemblies if the alkyl chain is longer than C_{10}. If one, or both, of the alkyl chains is shorter than C_8 micellar aggregates are formed. A micellar aggregate is distinguished from a bilayer by its lack of orderly molecular orientation. The hydrophilic head group may be replaced by anionic, nonionic or zwitterionic groups. The aggregate morphology is affected by the chemical structure, although the basic building block is commonly the bilayer assembly. As shown in Fig. 6.3, in the series $2C_n N^{\oplus} 2C_1$ [2] vesicle structure is predominantly observed for $n = 12$, but lamellae are the major morphology for $n = 18$. The layer width is 30~40 Å, a size consistent with that expected for a bilayer assembly. The molecular weights are extraordinarily large: $10^6 \sim 10^7$.

6.2.2   Bilayer formation by single-chain amphiphiles

It was mentioned above that the liquid crystalline nature is an intrinsic property of the bilayer membrane. Conversely, we may devise bilayer-forming compounds based on liquid crystalline substances, making use of the principle described in Fig. 6.4. p-Methoxybenzylidenebutylaniline (MBBA, [3]) is a typical, liquid crystal at room temperature. It is a rod-like molecule composed of a rigid segment (the diphenylazomethine moiety), which is planar and dipolar, and a flexible alkyl chain. Cetyltrimethylammonium bromide (CTAB, [4]) produces fluid globular micelles in water. A combination of the surfactant and liquid crystalline structures gives a bilayer-forming amphiphile [5]. This new type of amphiphile produces multiwalled vesicles when dispersed in water.

Single-chain amphiphiles with rigid segments give a wide variety of aggregate morphologies. It is concluded from an extensive survey by electron microscopy that aggregate morphology is determined by five structural elements of the amphiphile:

(1) flexible tail,
(2) rigid segment,
(3) hydrophilic head group,
(4) spacer group,
(5) additional interacting group.

The flexible tail is composed of linear methylene chains (C_7 or longer); and the rigid segment usually consists of two benzene rings (biphenyl, azobenzene, diphenylazomethine, etc.). The hydrophilic head group is common to those present in conventional surfactant molecules: e.g., ammonium, phosphate,

Fig. 6.3. Dialkylammonium bilayer membranes as observed by electron microscopy: (a) vesicle, (b) lamella.

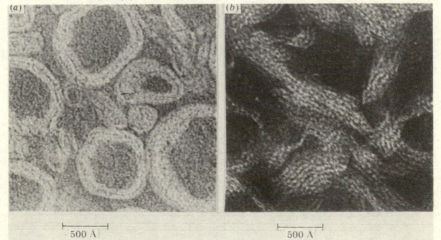

500 Å   500 Å

sulphate, phosphocholine and polyoxyethylene. The spacer group is the
methylene chain inserted between the rigid segment and the head group. The
additional interacting group may be an ester linkage near the tail end.

Representative aggregate morphologies are illustrated in Fig. 6.5. They
include globules, vesicles, lamellae, rods, tubes and disks. A globule is considered as a collection of less developed (or fragmentary) lamellae, and is different
from fluid, globular micelles. The two-dimensional development of the bilayer
assembly produces the typical bilayer membrane. The lamellar structure is
stacks of well-developed bilayers and the vesicle is made of closed bilayers. When
the bilayer structure possesses high curvature and develops uni-dimensionally,
rod-like and tubular structures will result. The rod-like structure is filled inside
and the tubular structure has the inner aqueous hole. A combination of flat
bilayers and portions of high curvature will produce disk-like structures. The
high-curvature portion constitutes the disk edge. The disk-like structures may
be isolated from each other or they may exist in stacked forms. The fundamental
structural unit of all the aggregate morphologies shown in Fig. 6.5 is the bilayer
assembly.

Fig. 6.4. Design of single-chain bilayer-forming compounds.

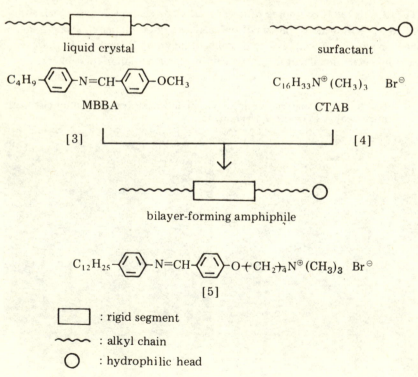

liquid crystal

surfactant

C_4H_9—N=CH—OCH$_3$

MBBA

$C_{16}H_{33}N^{\oplus}(CH_3)_3$ Br$^{\ominus}$

CTAB

[3]   [4]

bilayer-forming amphiphile

$C_{12}H_{25}$—N=CH—O$+CH_2)_4N^{\oplus}(CH_3)_3$ Br$^{\ominus}$

[5]

 : rigid segment

 : alkyl chain

 : hydrophilic head

The influence of the rigid segment on aggregate morphology is clearly seen in the electron micrographs of Fig. 6.6. These amphiphiles differ only in the rigid segment. The hydrophilic head group also affects the aggregate morphology. A single-chain ammonium amphiphile with the biphenyl rigid segment produces globular aggregates. In contrast, phosphate and phosphocholine head groups give fragmentary lamellae and stacked disks, respectively. The polyoxyethylene head groups give rise to rod-like structures. The trimethylammonium head group is least effective for aligning amphiphiles at the membrane surface.

6.2.3 *Monolayer formation by two-headed, single-chain amphiphiles*

Monolayer-forming compounds are two-headed amphiphiles which contain a rigid segment and a flexible chain. Examples are as follows:

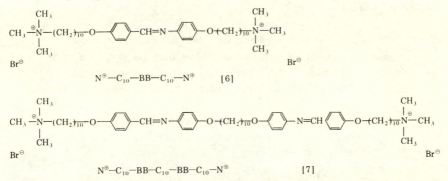

$$N^{\oplus}\!-\!C_{10}\!-\!BB\!-\!C_{10}\!-\!N^{\oplus} \quad [6]$$

$$N^{\oplus}\!-\!C_{10}\!-\!BB\!-\!C_{10}\!-\!BB\!-\!C_{10}\!-\!N^{\oplus} \qquad [7]$$

Fig. 6.5. Aggregate morphologies of single-chain amphiphiles.

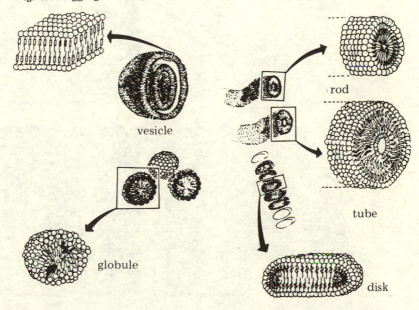

vesicle

rod

tube

globule

disk

The aggregate morphology of the monolayer membrane is influenced by structural factors that are different from those considered for the bilayer. In the case of $N^{\oplus}C_{10}$-X-$C_{10}N^{\oplus}$-type compounds in which the central rigid segments (x) are those of Fig. 6.6, the lamellar morphology shown in Fig. 6.7(a) is observed. The geometry of the rigid segment is not influential in this case. On the other hand, a two-headed amphiphile with a flexible chain in the middle of the molecule [7] gives a rod-like morphology (Fig. 6.7b). Examples of morphological changes that can be induced in monolayer membranes are illustrated in Fig. 6.8. Curvature is created for the lamellar $N^{\oplus}C_{10}BBC_{10}N^{\oplus}$ [6] by introducing an additional alkyl chain, cholesterol molecules, or a flexible methylene chain in the middle of the molecule. Depending on the extent of curvature (or molecular dissymmetry), multi-walled vesicles (small curvature) or rod-like structures (large curvature) will result.

6.2.4 *Membrane formation by polymers and polymerized vesicles*

Modification of the morphology of the bilayer aggregate is made possible by combination with polymers. Addition of polyanions such as poly-acrylate [8] causes destruction of the bilayer structure of dialkylammonium

Fig. 6.6. Electron micrographs of ammonium aggregates.

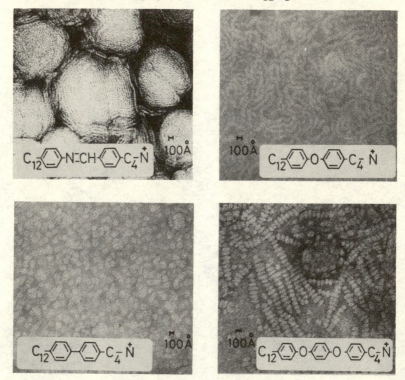

salts ($2C_nN^{\oplus}2C_1$) due to formation of polyion complexes and the subsequent precipitation. The influence of polyanions becomes moderate as the charge density along the polymer chain decreases. When copolymers of acrylate and acrylamide ([9] in Fig. 6.9) are added, stable aqueous solutions are obtained and the copolymer is inserted in the hydrophilic phase between the bilayers with concomitant bilayer separation, as shown in Fig. 6.9(a).

Fig. 6.7. Morphologies of monolayer aggregates: (a) lamella ($^{\oplus}NC_{10}B$-$BC_{10}N^{\oplus}$); (b) tube ($^{\oplus}NC_{10}BBC_{10}BBC_{10}N^{\oplus}$).

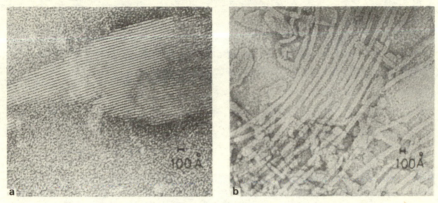

Fig. 6.8. Creation of curvatures in the monolayer membrane.

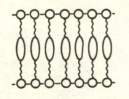

$N^{\oplus}-C_{10}-\bigcirc-C_{10}-N^{\oplus}$

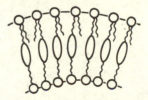

$N^{\oplus}-C_{10}-\bigcirc-C_{10}-N^{\oplus}$
$\diagdown C_{10}$

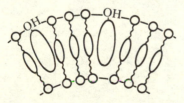

$N^{\oplus}-C_{10}-\bigcirc-C_{10}-N^{\oplus}$

+ cholesterol

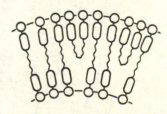

$N^{\oplus}-\bigcirc-C_{10}-\bigcirc-N^{\oplus}$

$\bigcirc- \quad ; -\langle\_\rangle-N=CH-\langle\_\rangle-$

Complexation of polymers with the bilayer may be attained by incorporation of polymer side chains into the bilayer matrix. Water-soluble polymers with monoalkyl or dialkyl side chains ([10] and [11] in Fig. 6.9) do not form definite aggregate structures when dispersed in water. When $C_{12}AAm$-AAm (20 wt %) is mixed with the $2C_{12}N^{\oplus}2C_1$ ([2], $n = 12$) membrane, the bilayer structure is destroyed progressively with increasing content of the alkyl side chain in the polymer. Apparently, the single alkyl side chain, when incorporated, destroys the bilayer assembly, just like single-chain surfactants (CTAB [4] and others). On the other hand, the bilayer structure is well retained in the case of a copolymer with 5~10 mol % of a dialkyl side chain. A schematic illustration of the structure of this membrane–polymer complex is shown in Fig. 6.9(b).

These results suggest that the bilayer assembly may be obtained from water-soluble polymers alone. Among the various combinations of monomers, the following copolymers give well-developed bilayer vesicles.

Fig. 6.9. Membrane–polymer complexes.

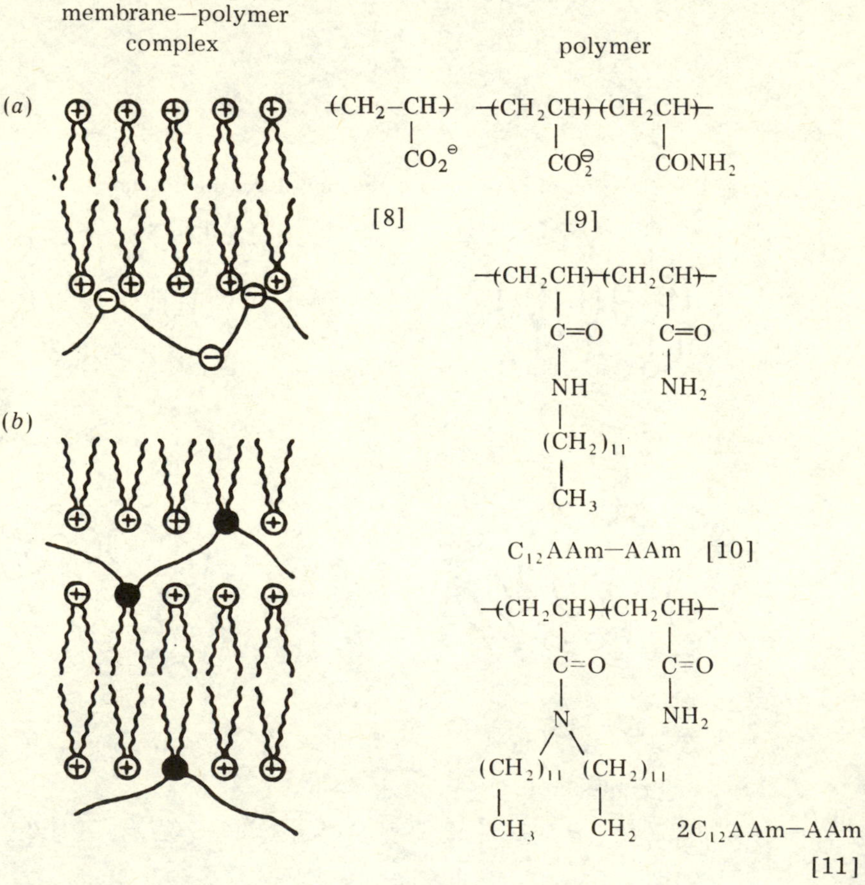

membrane—polymer
complex   polymer

(a)

$+CH_2$–$CH+$
    |
   $CO_2^{\ominus}$

$-(CH_2CH)(CH_2CH)-$
        | |
       $CO_2^{\ominus}$ $CONH_2$

[8]   [9]

$-(CH_2CH)(CH_2CH)-$
        | |
       $C=O$ $C=O$
        | |
        NH NH_2
        |
     $(CH_2)_{11}$
        |
       CH_3

(b)

$C_{12}AAm$—AAm [10]

$-(CH_2CH)(CH_2CH)-$
        | |
       $C=O$ $C=O$
        | |
        N NH_2
       / \
 $(CH_2)_{11}$ $(CH_2)_{11}$
      | |
     CH_3 CH_2 $2C_{12}AAm$—AAm

[11]

$$\fbox{CH_2{-}CH}_{100-x}\fbox{CH_2{-}CH}_x \qquad n = 12, 16$$
$$\begin{array}{cc} R_1 & R_2 \end{array} \qquad x = 7\text{–}15$$

[12] : R_1 = —CNH_2; R_2 = —$\underset{\underset{O}{\|}}{C}$—NH—$\underset{\underset{(CH_2)_{\overline{7}}CO{-}(CH_2)_{\overline{n-1}}CH_3}{\underset{\|}{O}}}{\overset{\overset{H}{|}}{C}}$—$\overset{\overset{O}{\|}}{C}$—O—$(CH_2)_{\overline{n-1}}CH_3$

[13] : R_1 = —CNH_2; R_2 = —$\underset{\underset{O}{\|}}{C}$—$(OCH_2CH_2)_{\overline{4}}O\underset{\underset{O}{\|}}{C}$—$CH_2CH_2$—$\overset{\overset{H}{|}}{C}N$—$\overset{\overset{H}{|}}{\underset{\underset{(CH_2)_{\overline{7}}CO(CH_2)_{\overline{n-1}}CH_3}{\underset{\|}{O}}}{C}}$—$\overset{\overset{O}{\|}}{C}$—O

It is evident that the vesicles are composed of the bilayer structure, such as shown by the schematic illustration of Fig. 6.10(*a*). The copolymer vesicle is capable of retaining water-soluble substances in the inner water core. Glucosamine is a useful probe which can be detected after its conversion to a fluorescent compound by reaction with fluorescamine: cf. §6.5.4. The extent of retention is 2~3% for the vesicles and very small for lamellae of the copolymer.

A molecular membrane can also be formed by folding of the polymer main chain. Simple ionenes are prepared by the following procedure:

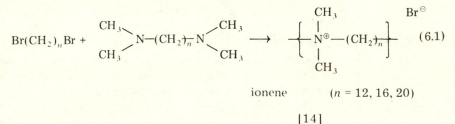

$$Br(CH_2)_n Br + \underset{CH_3}{\overset{CH_3}{>}}N{-}(CH_2)_n{N}\underset{CH_3}{\overset{CH_3}{<}} \longrightarrow \left[\begin{matrix} CH_3 \\ | \\ N^\oplus{-}(CH_2)_n \\ | \\ CH_3 \end{matrix}\right] \quad Br^\ominus \qquad (6.1)$$

ionene   (*n* = 12, 16, 20)

[14]

Fig. 6.10. Polymeric molecular membranes: (*a*) polymer bilayer, (*b*) ionene monolayer.

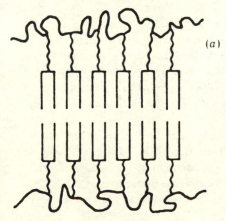

(*a*)

(*b*)

The polymers give clear dispersions upon sonication. Electron microscopy indicates that vesicles are formed from inonene ($n = 12, 20$) and that vesicles and lamellae are present for aqueous inonene ($n = 16$).

The bilayer structure is not clearly seen by electron microscopy for either the monomeric or polymerized aggregates of the following compounds:

$$\begin{array}{l}
CH_2{=}C{\diagup}^{R} \\
\end{array}$$

$$CH_2{=}C\overset{\displaystyle R}{\diagdown}\;\underset{\displaystyle O}{\overset{\displaystyle \|}{C}}-O-(CH_2)_n\diagdown\;\underset{C_mH_{2m+1}}{\overset{}{N^{\oplus}}}\diagup\overset{\displaystyle CH_3}{\underset{\displaystyle CH_3}{}}$$

$R = H$ ($n, m = 10, 12; 10, 18; 16, 12; 16, 18$)

$R = CH_3$ ($n, m = 16, 18$)

$$Br^{\ominus}$$

[15]

However, well-defined vesicles are obtainable when the ammonium group is replaced by the phosphocholine group. Attempts to stabilize bilayer vesicles by polymerization were also reported by three independent research groups almost simultaneously. They coincidentally synthesized dialkyl amphiphiles with the

$$C_nH_{2n+1}-C{\equiv}C-C{\equiv}C-(CH_2)_8-\overset{\displaystyle O}{\overset{\displaystyle \|}{C}}-O-CH_2$$

$$C_nH_{2n+1}-C{\equiv}C-C{\equiv}C-(CH_2)_8-\underset{\displaystyle O}{\overset{\displaystyle \|}{C}}-O-CH$$

$$CH_2-O-\underset{\displaystyle O^{\ominus}}{\overset{\displaystyle O}{\overset{\displaystyle \|}{P}}}-O-CH_2-CH_2-N^{\oplus}-(CH_3)_3$$

[16]

$n = 9, 11, 13$

diacetylene moiety in the centre of the alkyl chain. The monomer vesicle readily polymerized upon irradiation, displaying interesting colour changes that are typical of the diacetylene compound. The ease of polymerization is highly sensitive to the monomer packing and extensive conjugation is possible in the vesicle. The polymerized vesicle shows enhanced stability.

6.3 Physicochemical characteristics

6.3.1 *Phase transition and spectral change*

The synthetic bilayer membrane possesses physicochemical characteristics peculiar to the bilayer assembly. Some of these properties are described here for the dialkylammonium bilayer as an example.

The NMR spectrum of the CTAB [4] micelle is composed of sharp peaks, indicating that the CTAB molecule is very fluid in the micelle. In contrast, the spectrum of the $2C_{18}N^{\oplus}2C_1$ ($n = 18$ in [2]) bilayer is quite broad, suggesting that the molecular motion is restricted. The peaks of $C_{18}C_8N^{\oplus}2C_1$, which is incapable of bilayer formation, are somewhat broad but close in shape to those of the CTAB micelle.

A drastic change in the peak width is observed in the NMR spectrum and is ascribed to the gel-to-liquid crystal phase transition of the bilayer. This is confirmed by heat absorption due to phase transition as measured using differential scanning calorimetry. The samples are dilute aqueous dispersions of $2C_nN^{\oplus}2C_1$ ([2], $n = 12, 14, 16,$ and 18) obtained by sonication at room temperature and those dispersions kept at $-50\,^\circ$C. The former samples give broader peaks than the latter at low temperatures. The phase transition temperature (T_c) rises with increasing alkyl chain length in the membrane, as summarized in Table 6.1.

Similar thermal behaviour is observed for other dialkyl amphiphiles with anionic, nonionic and zwitterionic head groups. The overall trend, such as the dependence of T_c on the alkyl chain length, is similar to that of biolipids. A schematic illustration of the phase transition is also shown in the table. The alkyl chain is crystallized by assuming the extended *trans* conformation at temperatures below T_c, but turns liquid crystalline by melting at higher temperatures.

The molecular orientation in the bilayer matrix leads to unique spectral characteristics for chromophore-containing amphiphiles.

$$CH_3(CH_2)_{11}OC-C^*-N-C-\underset{}{\bigcirc}-O-(CH_2)_n-\overset{\oplus}{N}-CH_3 \quad Br^{\ominus} \quad [17]$$

$$2C_{12}GluPhC_nN^{\oplus} \qquad n = 2, 4, 6$$

For example, amphiphile [17] gives well-organized bilayer vesicles. This compound contains a chiral centre, and shows enhanced optical activity upon

Table 6.1. *Phase transition temperature (T_c) of dialkylammonium bilayer membranes*

Bilayer	T_c (°C)	
	Sonicated sample	Frozen sample ($-50\,^\circ$C)
$2C_{18}N^{\oplus}2C_1$	45	54
$2C_{16}N^{\oplus}2C_1$	28	44
$2C_{14}N^{\oplus}2C_1$	16	31
$2C_{12}N^{\oplus}2C_1$	—	16

(*a*) crystalline (gel) phase
(*b*) liquid-crystalline phase

formation of the bilayer. Fig. 6.11 shows the temperature dependence of the circular dichroism (CD) spectrum of this bilayer in dilute aqueous solution (1.0×10^{-4} M). At temperatures above $31 \sim 32\,^{\circ}$C (T_c), the CD spectrum possesses a maximum at 245 nm with $[\theta]_{245} = +6000$. This spectrum is identical with that observed in methanol. At temperatures below T_c, one observes extremely large maxima at 220 and 260 nm with a shoulder extending beyond 700 nm: $[\theta]_{220} = +360\,000$ and $[\theta]_{260} = -400\,000$. The enhanced optical activity is lost upon lowering of the surfactant concentration below the critical micelle concentration (CMC), and upon addition of bilayer-destroying compounds such as CTAB ([4]). The enormously enhanced circular dichroism is derived from the orientational fixation of chiral dialkyl amphiphiles in the rigid bilayer assembly.

$$CH_3 \diagdown N-\langle\bigcirc\rangle-N{=}N-\langle\bigcirc\rangle-SO_3{}^{\ominus}Na^{\oplus} \qquad \text{methyl orange [18]}$$
$$CH_3 \diagup$$

Strong CD was induced with methyl orange bound to the chiral membrane. The induced circular dichroism (ICD) spectrum is extremely sensitive to the phase transition of the matrix membrane. The spectrum is invariable up to $26 \sim 27\,^{\circ}$C with a large θ_{max} value of 200 000, but disappears completely at

Fig. 6.11. CD spectra of chiral bilayers in water. The concentration of L-[17] $= 1.0 \times 10^{-4}$ M.

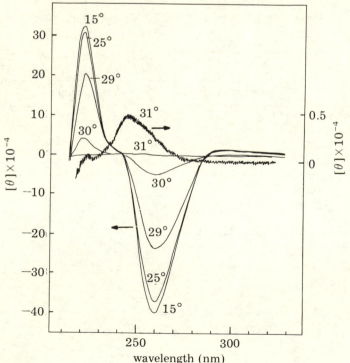

29.5 °C, reflecting the fluidity change of the membrane at T_c (31~32 °C). Thus, as in the above CD case (Fig 6.11), the orientational fixation of the chromophore (bound methyl orange) on the rigid chiral bilayer is crucial for the ICD observation.

The spectral control by the bilayer matrix is not limited to CD and ICD spectra. The visible spectrum of bound methyl orange is also highly sensitive to minor variations in the amphiphile structure and to the molar ratio of dye to amphiphile. The ease of dye aggregation at the membrane surface appears to be influenced by membrane fluidity.

The change in the binding mode of methyl orange due to phase transition is schematically illustrated in Fig. 6.12.

6.3.2　*Phase separation*

When the bilayer is composed of more than one component, they are not necessarily miscible. The inhomogeneity – phase separation – is known to be an important factor for controlling the physiological function of the biomembrane.

The phase separation in the synthetic bilayer membrane is detected by thermal measurement (differential scanning calorimetry) or by absorption spectroscopy. A single-chain amphiphile with the azobenzene rigid segment,

Fig. 6.12. Schematic illustration of methyl orange bound to the bilayer membrane.

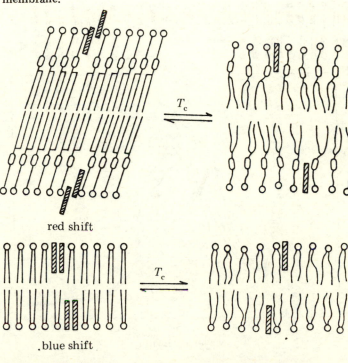

red shift

.blue shift

$C_{12}AzoC_nN^{\oplus}$ [19], produces globular aggregates (diameter: 200~300 Å) made of lamellar fragments.

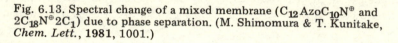

$$CH_3(CH_2)_{11}O-\langle\bigcirc\rangle-N=N-\langle\bigcirc\rangle-O-(CH_2)_nN^{\oplus}-CH_3 \quad Br^{\ominus}$$

$$C_{12}AzoC_nN^{\oplus} \qquad n = 2, 4, 10 \qquad [19]$$

The absorption spectrum of [19] possesses an absorption maximum (λ_{max}) at 355 nm in ethanol and at 330 nm in water. It is well known that azo dyes absorb at shorter wavelengths upon aggregation. Therefore, the spectral data of [19] in water are caused by stacking of the chromophore.

The spectrum of [19] in the $2C_nN^{\oplus}2C_1$ [2] bilayer matrix shows temperature dependence. Fig. 6.13 gives the spectra of a 1:10 mixture of [19] and $2C_{18}N^{\oplus}2C_1$ [2]. The absorption maximum is located at 355 nm at 50 °C, implying that the azobenzene chromophore is in the monomeric form. By

Fig. 6.13. Spectral change of a mixed membrane ($C_{12}AzoC_{10}N^{\oplus}$ and $2C_{18}N^{\oplus}2C_1$) due to phase separation. (M. Shimomura & T. Kunitake, *Chem. Lett.*, **1981**, 1001.)

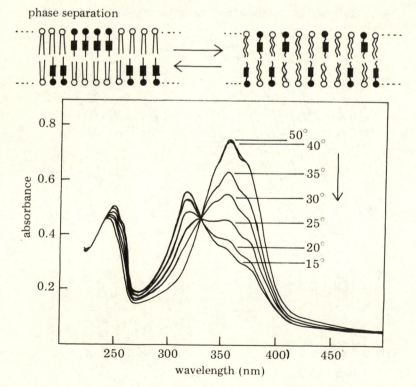

lowering the temperature, the peak height at 355 nm diminishes, and a new peak due to the stacked chromophore appears at 320 nm. It is thus suggested that the azobenzene chromophore is miscible with the $2C_{18}N^{\oplus}2C_1$ bilayer matrix at a high temperature range but that phase separation occurs at low temperature (see illustration at top of Fig. 6.13). The phase separation is apparently governed by the fluidity of the matrix membrane, and the azobenzene amphiphile separates as the matrix membrane crystallizes at T_c (45 °C for $2C_{18}N^{\oplus}2C_1$).

6.4 Reaction control
6.4.1 *Specific activation of a cholesteryl nucleophile*

Cholesterol is contained in large amounts in the plasma membrane of animals and plays an important role in the membrane function. The synthetic bilayer membrane of dialkyl amphiphiles can incorporate cholesterol, and the bilayer structure is not destroyed by 20~30% of cholesterol. Conventional surfactants, if added in the same amount, destroy the membrane structure. Thus, the hydrophobic cholesterol skeleton appears to be bound snugly in the membrane matrix.

It is known that cationic micelles activate bound anionic reagents. If this activation is combined with the specific binding of anions to the membrane, a new catalytic system, both powerful and specific, may be achieved. Table 6.2 summarizes the catalytic hydrolysis of *p*-nitrophenyl acetate (PNPA) by nucleophiles in cationic matrices.

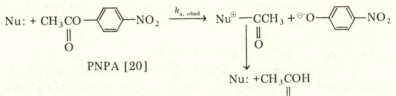

PNPA [20]

nucleophilic catalyst

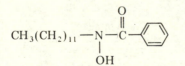

C_{12}—BHA [21]

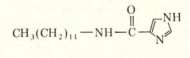

C_{12}—ImAm [22]

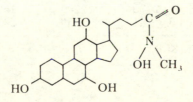

chol-HA [23]

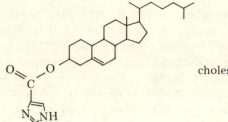

cholest-Im [24]

The enhanced reactivity of non-specific, long-chain nucleophiles, C_{12}-BHA [21] and C_{12}-ImAm [22], may be explained in terms of a 'hydrophobic ion pair': formation of less solvated (i.e. activated) anions by pairing with hydrophobic countercations. The matrix effect changes in the order of TMAC $> 2C_{18}N^{\oplus}2C_1 > 2C_{12}N^{\oplus}2C_1 >$ CTAB. In the case of chol-HA [23] (the effective nucleophile is the dissociated species), activation by CTAB is greater than that of $2C_{12}N^{\oplus}2C_1$. In contrast, cholest-Im [24] is activated about 40 times more efficiently in the $2C_{12}N^{\oplus}2C_1$ membrane than in the CTAB micelle. Furthermore, whereas the bilayer-bound cholest-Im is more active than that bound to TMAC, the extent of activation is reversed for the other nucleophiles. These results suggest that membrane activation of the cholest-Im nucleophile is attributed not to the simple hydrophobic effect but to the specific binding to the membrane.

Cholic acid is known to destroy the phospholipid bilayer. In the same vein, addition of 30% cholic acid was shown by electron microscopy to destroy the lamellar structure of $2C_{18}N^{\oplus}2C_1$. Cholic acid derivatives like chol-HA [23], are not specifically activated by the bilayer membrane, which is consistent with the electron microscopy results.

6.4.2 *Membrane fluidity and catalysis*

Reactions occurring in the bilayer matrix are affected by the matrix fluidity, as one would expect. The decarboxylation reaction of equation 6.2 is a unimolecular reaction that is devoid of concentration dependence and is highly

Table 6.2. *Rate of nucleophilic cleavage of PNPA*, $k_{a.obsd}$ $(M^{-1}s^{-1})$

		Aggregate			
Nucleophile	None	CTAB (1×10^{-3} M)	TMAC (1×10^{-4} M)	$2C_{12}N^{\oplus}2C_1$ (1×10^{-3} M)	$2C_{18}N^{\oplus}2C_1$ (1×10^{-3} M)
C_{12}-BHA [21]	15	1900	6700	4500	—
C_{12}-ImAm [22]	0.09	100	1200	160	810
chol-HA [23]	10	2770	2500	1600	2800
cholest-Im [24]	0.09	61	650	2600	—

30 °C, pH 8.9, $\mu = 0.01$ (KCl).
[PNPA] $= 3.79 \times 10^{-6}$ M.

sensitive to the microenvironment. Therefore, this reaction has been often used as a kinetic probe of the microenvironment.

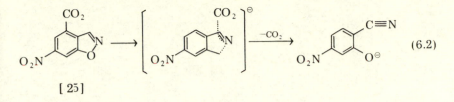

$$\qquad\qquad\qquad\qquad\qquad\qquad\qquad\qquad\qquad\qquad (6.2)$$

[25]

The reaction rate in the ammonium bilayer matrix is affected by the hydrophobicity and fluidity of the matrix. The Arrhenius plots of this reaction are expressed by two lines with breaks near the phase transition temperature. There is no break in the CTAB micelle system. The apparent activation energy is estimated from the slope, and is given in Table 6.3. In the temperature range above T_c, the activation energy decreases with increasing alkyl chain length in the matrix. By contrast, it increases with increasing chain length in the temperature range below T_c. These results suggest that the reaction is accelerated in the fluid membrane matrix by the matrix hydrophobicity but that the lack of fluidity is more influential than hydrophobicity for the membrane matrix in the gel state.

Similar anomalies in the Arrhenius plots are observed for nucleophilic displacement and proton abstraction. In all cases, the activation energy is greater at temperatures above T_c than below T_c.

6.4.3   *Reaction control by specific orientation and distribution*

The alkyl chain is highly oriented in the bilayer membrane. This characteristic may be used to construct new molecular devices which are not feasible with conventional surfactant micelles.

The reaction control that can be realized by taking advantage of the difference in the location of a reacting moiety in the membrane matrix is such an example.

Table 6.3. *Activation energy of decarboxylation*

Ammonium aggregate	E_a (kcal mol^{-1})	
	Below T_c	Above T_c
CTAB micelle		(20)
$2C_{12}N^{\oplus}2C_1$ membrane	18	25
$2C_{14}N^{\oplus}2C_1$ membrane	22	20
$2C_{16}N^{\oplus}2C_1$ membrane	25	19
$2C_{18}N^{\oplus}2C_1$ membrane	28	14

The following compounds form the bilayer assembly:

$$CH_3(CH_2)_{11}-\!\!\left\langle\!\!\bigcirc\!\!\right\rangle\!\!-N=CH-\!\!\left\langle\!\!\bigcirc\!\!\right\rangle\!\!-\overset{\overset{\displaystyle CH_3}{|}}{\underset{\underset{\displaystyle CH_3}{|}}{N^{\oplus}}}\!\!-CH_3 \quad Br^{\ominus}$$

$$C_{12}BBC_0N^{\oplus} \qquad [26]$$

$$CH_3(CH_2)_{11}\!\!\left\langle\!\!\bigcirc\!\!\right\rangle\!\!-N=CH-\!\!\left\langle\!\!\bigcirc\!\!\right\rangle\!\!-O-(CH_2)_{\overline{m}}\overset{\overset{\displaystyle CH_3}{|}}{\underset{\underset{\displaystyle CH_3}{|}}{N^{\oplus}}}\!\!-CH_3 \quad Br^{\ominus}$$

$$C_{12}BBC_m N^{\oplus} (m = 4, 10) \qquad [27]$$

The azomethine group in the rigid segment reacts readily with water and amines. Its reactivity may be influenced by its distance from the membrane surface. Under weakly acidic conditions (pH 5.5, 30 °C), the reaction with water decreases with increasing spacer length (C_m), the rate for C_{10} being one fiftieth of that for C_0. In the reaction with amines (Fig. 6.14), the reaction rate is not particularly sensitive to C_m when benzylamine is used. Benzylamine is considered to be miscible with the membrane matrix and the distance of the azomethine group from the membrane surface is not influential. Dodecylamine may be oriented in the membrane and the amine group will be located near the membrane surface, thus larger rate differences being observed. Polyethylenimine [28] is water-soluble and cannot penetrate into the hydrophobic membrane matrix. As expected, reaction is normal in the absence of the spacer, but is

Fig. 6.14. Reaction of amines with $C_{12}BBC_m N^{\oplus}$ bilayers (30 °C, pH 6.0).

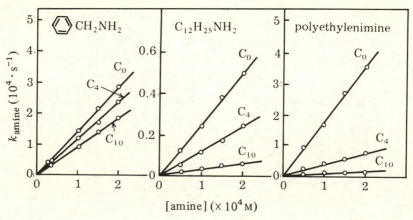

suppressed appreciably in the case of the C_4 and C_{10} spacers.

$$\text{~N~NH~}$$
$$\underset{NH_2}{|}$$

polyethylenimine

[28]

Phase separation is a useful technique for reaction control in the bilayer matrix. An ester substrate with the azobenzene chromophore $C_{12}AzoOCOC_{10}N^{\oplus}$ [29] forms the bilayer assembly upon dispersion in water. A 1:10 aqueous mixture of this substrate and $2C_nN^{\oplus}2C_1$ [2] gives absorption maxima at $310\sim315$ nm at 10 °C, and 350 nm at 50 °C. Thus the substrate exists as clusters at 0 °C and as isolated species at 50 °C. The rate of alkaline hydrolysis is affected by the formation of the substrate cluster. The anomaly in the Arrhenius plots is related to phase separation, the reaction rate decreasing upon phase separation at low temperatures.

$$CH_3(CH_2)_{11}-O-\!\!\!\!\langle\rangle\!\!\!\!-N=N-\!\!\!\!\langle\rangle\!\!\!\!-O\underset{\underset{O}{\|}}{C}-(CH_2)_{10}-\underset{\underset{CH_3}{|}}{\overset{\overset{CH_3}{|}}{N^{\oplus}}}-CH_3 \quad Br^{\ominus}$$

$$C_{12}AzoOCOC_{10}N^{\oplus}$$

[29]

6.4.4 *Substrate entrapment and inter-vesicle reaction*

The synthetic bilayer membrane readily forms single- and multi-walled vesicles (see Fig. 6.2). These vesicles can trap water-soluble substances in their inner water cores. Trapping of sparingly water-soluble substances in the bilayer matrix is also possible.

Fig. 6.15 describes a binding experiment of methyl orange to several ammonium aggregates. When methyl orange bound to the $2C_{12}N^{\oplus}2C_1$ and $2C_{18}N^{\oplus}2C_1$ membrane is dialysed against water, little leakage is observed even after 100 h. Dialysis proceeds smoothly with the CTAB micelle or the TMAC aggregates. The bilayer formation results in remarkably reinforced binding.

Fendler *et al.* carried out the binding experiment with the $2C_{18}N^{\oplus}2C_1$ bilayer aggregate. The retention of hydrophilic amino acids (L-alanine and L-serine) is not efficient, but hydrophobic dansyl-*n*-octadecylamine is more efficiently trapped. This amphiphile produces predominantly lamellar aggregates. Therefore, substrates may be trapped in the interlamellar phase or in the membrane matrix.

Substrate trapping in the inner water core is more clearly shown by the following experiment. Highly water-soluble amines such as glucosamine [30] or ethanolamine are not trapped at the vesicle surface or in the membrane matrix. Vesicles that trap these amines are separated from amines in the bulk water phase by gel chromatography, and the trapped amines are allowed to react with fluorescamine for detection. The bilayer-forming amphiphile [17] gives double-

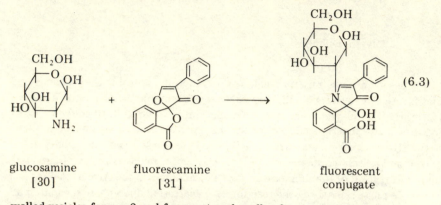

glucosamine
[30]

fluorescamine
[31]

fluorescent
conjugate

(6.3)

walled vesicles for $n = 2$ and fragmentary lamellae for $n = 4$. The trapping efficiency of glucosamine [30] by the vesicles is 2.5%, whereas that by the lamellae is negligibly small ($\sim 0\%$). Since the molecular structures of the component amphiphiles are very similar, the difference in the trapping efficiency

Fig. 6.15. Leakage of methyl orange ([18]) bound to ammonium aggregates. ○: $2C_{18}N^{\oplus}2C_1$ membrane, ●: $2C_{12}N^{\oplus}2C_1$ membrane, △: CTAB micelle, □: TMAC aggregate. (Y. Okahata, R. Ando & T. Kunitake, *Bull. Chem. Soc. Japan*, **52**, 3647 (1979).)

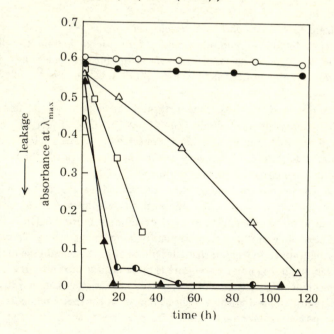

must come from the change in aggregate morphology. The experiments show that hydrophilic substances are not readily permeable across the hydrophobic bilayer. It was found recently that even $OH^{\ominus}$ cannot freely pass the ammonium bilayer membrane. This leads to formation of a pH gradient across the membrane, the most important membrane property in bioenergetics (see Chapter 5).

Reaction control by substrate entrapment in vesicles is possible. As shown in Fig. 6.16, when catalyst (C) and substrate (S) are trapped in separate vesicles, either reagent must be transferred prior to reaction. Thus, if the rate of reagent transfer is slow, the intervesicle reaction would be slower than the intra-vesicle reaction.

Table 6.4 compares the apparent rate constants (k_{obsd}) for the series A experiment in which catalyst (cholest-Im [24]) and *p*-nitrophenyl esters [32] are retained in separate ammonium bilayer aggregates and the series B experiment in which catalyst and substrate are retained in the same aggregate. In the case of PNPA substrate ([32], $R = CH_3$), the reaction rates are almost the same between the two series of experiments in the micellar and bilayer systems. This indicates that PNPA substrate is rapidly transferred among the vesicles. When the more hydrophobic PNPP substrate ([32], $R = C_{15}H_{31}$) is used, the k_{obsd} value in the fluid CTAB micelle is the same between the two series. However, k_{obsd} is different by a factor of more than 200 in the $2C_{12}N^{\oplus}2C_1$ membrane system, suggesting that transfer of hydrophobic PNPP is the slow step. The separate ageing experiment supports this conclusion, and the rate constant of substrate transfer is

Fig. 6.16

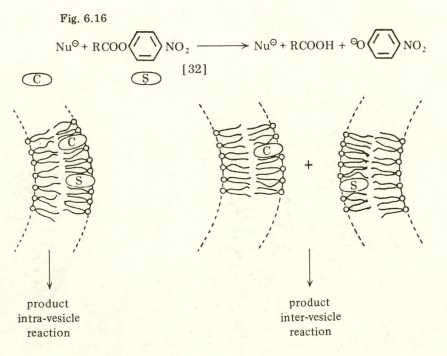

$$Nu^{\ominus} + RCOO\!\!-\!\!\langle \; \rangle\!\!-\!\!NO_2 \longrightarrow Nu^{\ominus} + RCOOH + {}^{\ominus}O\!\!-\!\!\langle \; \rangle\!\!-\!\!NO_2$$

[32]

product
intra-vesicle
reaction

+

product
inter-vesicle
reaction

estimated as 1×10^{-4} s^{-1}. This is very small relative to that in the conventional micellar system ($10^3 \sim 10^7$ s^{-1}).

The process of the inter-vesicle reaction can be examined more closely by taking advantage of the spectral change of azobenzene-containing amphiphiles. Aqueous bilayer aggregates of C_{12}-Azo-C_{10}-N$^\oplus$ [19] and $2C_{18}$N$^\oplus$2C$_1$ [2] are prepared separately and mixed. The azobenzene amphiphile initially exists solely as the cluster species (see §6.3.2). The isolated azobenzene chromophore is produced gradually upon mixing of the two solutions. A schematic illustration of the fusion process is shown in Fig. 6.17.

6.5 Photochemistry

The use of the synthetic bilayer membrane as a site of photochemical events is rapidly expanding. This is because of the realization that the bilayer assembly is an intrinsic part of the photosynthetic machinery in the biological system. The conversion of light to chemical energy is discussed in Chapter 4, and will not be repeated here. An artificial light-energy converter in the vesicle system has been proposed by Matsuo (1980). The bilayer characteristics are essential for constructing this system.

Table 6.4. *Rates of inter- and intra-vesicle reactions*

	k_{obsd} (s^{-1})			
	$2C_{12}$N$^\oplus$2C$_1$ bilayer		CTAB micelle	
	PNPA	PNPP	PNPA	PNPP
inter-vesicle	0.53	0.032	0.066	0.12
intra-vesicle	0.45	7.8	0.065	0.19

30 °C, pH 9.5, $\mu = 0.01$.

Fig. 6.17

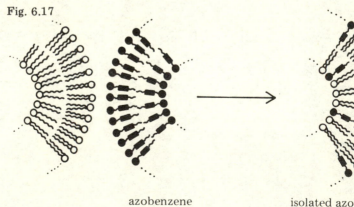

azobenzene
cluster

isolated azobenzene
chromophore

7

INFORMATION-TRANSMITTING MACROMOLECULES

IWAO TABUSHI & MORIO IKEHARA

7.1 General introduction

Today much information floods the world from radio, television, magazines, and newspapers. However, Nature always has lived, and will continue to live, without this created information. This chapter discusses information transmittance not of the artificial type, but of information that is reproduced, copied and transmitted in native bio-systems. All the necessary information (at least of life in its present form) is recorded in nucleic acids and transmitted as a fundamental (entropy) flow, while proteins synthesized according to this information maintain the practical functions of living systems. DNA and RNA are the information-transmitting macromolecules and the main theme of this chapter concerns their synthesis and properties. Information carried by nucleic acids is close to a purely numerical (statistical) type which is determined by nucleotide sequences. However, information governing real living systems is transmitted by proteins such as signal substances, functional substances (receptors, enzymes, carriers) and memory substances. Though proteins accept the information of amino acid sequence (primary structure) from nucleic acids, proteins utilize pattern information in their functions due to their structure and nature (e.g., charge distribution, coordination, hydrogen bonding, hydrophobic groups). This chapter briefly discusses information-transmitting macromolecules carrying pattern information (i.e., proteins).

The information transmitted as DNA → mRNA → protein is defined by three nucleotide units which code for one amino acid unit. In this sequence determination of a peptide, information corresponding to a probability of $p = (1/4)^3 = 1/64$ is required for the selection of a single amino acid unit, the information of which corresponds to only the probability of $p = 1/20$. Therefore, information quantity clearly decreases. In spite of this decrease, Nature still utilizes proteins for practical information flow, probably because of their variety. Amino acid side chains in proteins have various physical properties: hydrophobic groups like phenyl or alkyl, negatively charged groups like $-COO^{\ominus}$, positively charged groups like or $-NH_3^{\oplus}$, metal-chelating residues like imidazole, $-CO_2^{\ominus}$ or $-NH_2$, bridging groups like $-SH$. When the amino acid sequence (i.e. information from DNA) has been given, the overall structure of the polypeptides is determined by mutual interaction among the residues in order to gain the most stable or metastable states. Surfaces and cavities might be constructed specifically by gathering amino acid residues. Examples may be seen in the construction of a hydrophobic cavity of a defined shape and size, chelating residues for metal incorporation, and assembly of the same ionic groups on the surface for assembly recognition. These specific sites interact specifically and very rapidly with another protein's complementary sites, i.e. with another hydrophobic surface, metal ion or assembly of counter ions on the surface. Therefore, recognition may occur at multiple points and the information is of the pattern-recognition type.

This pattern information is probably used because it allows very rapid and specific information transmittance. Association rate constants between proteins and functional molecules often reach $10^7 \, \mathrm{s}^{-1} \, \mathrm{M}^{-1}$ and association constants (specific) are quite large. For instance, the enzyme trypsin and its specific inhibitor associate with an association constant of $10^{14} \, \mathrm{M}^{-1}$, and for the hormone insulin and the insulin receptor it is $10^{11} \, \mathrm{M}^{-1}$.

Moreover, proteins are much more portable than nucleotides, since in proteins three units of nucleotides are replaced by one amino acid, leading to a reduction in weight to about a tenth that of the nucleic acid. Also, as polypeptides are usually folded in a rather compact form, they may be much more easily transported than nucleic acids. However, the polypeptide information is a pattern-recognition type and at present is rather difficult to understand. The numerical information of nucleic acids is much easier to analyse than the information of peptides, and rapid progress in nucleic acid chemistry makes it easier to understand the mechanism of the information transmittance. This chapter discusses mostly the chemical synthesis of nucleic acids which are considered the most important information-transmitting macromolecules.

We point out here that numerical information sources are not restricted to nucleic acids. A macromolecule which has optimum properties of interaction for recognition will act, in principle, as an information transmitter, although it is not as accurate or effective. Nucleotides can carry a high degree of information based on the 'one out of four' principle, namely selectivity between double and triple hydrogen bonding and that between donor and acceptor character. If the simplest chemical interaction, which is determined by the 'all or none' principle, is satisfied, almost all chemical interactions must be utilized in a mutual recognition interaction. This will lead to innumerable possibilities for utilizing information-transmitting artificial macromolecules. The most primitive form is the so-called matrix polymerization, where only one kind of recognition interaction (not sequential information) constitutes the system and the kind and number (equal to the number of an information source) of

the counterpart is determined. In this special example, the most probable
degree of polymerization is determined. For instance, electrostatic interaction (Fig. 7.1) or hydrogen bonding interaction can act as the recognition
interaction. The degrees of polymerization of the information source will be
transmitted to that of the products. Thus, the minimum information will be
transmitted by simple chemical interaction through simply constructed synthetic
macromolecules. On this basis, complex combination of the type and number of
the elemental interaction may provide sequential or other sophisticated information, which will lead to substances rapidly carrying information as well as
appropriate functions. We should then be able to prepare pure synthetic
macromolecules transmitting information capable of microscopic (at the atomic
or molecular level) control of a certain reaction or phenomenon.

The last section of this chapter describes information transmittance in biological systems, especially genetic information related to nucleic acids.

7.2   Synthesis of nucleic acids

Nucleic acids are macromolecules of nucleosides consisting of hetero-
cyclic bases and ribose or 2-deoxyribose, connected by phosphodiester linkages
between 3'- and 5'-hydroxyl groups (Fig. 7.2). Their molecular weights are
between tens of thousands and several millions. They are synthesized in biological systems from nucleoside-5'-triphosphates. Synthesis is catalysed by enzymes
at rates of a few tens of nucleotides per second (in P2 phage). Their sequences
are almost always accurate, which attests to the precision of bioorganisms.

7.2.1   *Synthesis by the phosphodiester method*

In the early 1950s, A. R. Todd developed the phosphotriester method, as
it is now called, to synthesize a dinucleoside monophosphate (TPT) as the
simplest unit of nucleic acid. However, the benzyl group used in the phosphate
protection was too labile to acid, and the phosphodiester method replaced it.

Fig. 7.1. Interactions at active enzyme sites.

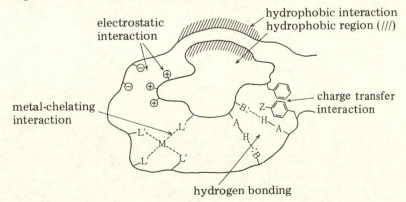

In the phosphodiester method, as shown in Fig. 7.3, nucleosides having a phosphate on 3′ or 5′-OH (3′- or 5′-nucleotides) are properly protected and condensed with nucleoside having 5′- or 3′-OH exposed for reaction using a certain condensing reagent. These two approaches are used in the ribonucleotides method (i) and the deoxyribonucleotides method (ii) because of the ease of obtaining the starting materials and the good protecting procedures.

Fig. 7.2. Structure of nucleic acids.

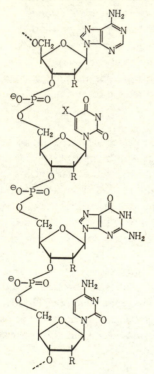

DNA (R = H)

Base	*Nucleoside*
adenine (A)	2′-deoxy-adenosine
thymine (T) (X = CH$_3$)	2′-deoxy-thymidine
guanine (G)	2′-deoxy-guanosine
cytosine (C)	2′-deoxy-cytidine

RNA (R = OH)

Base	*Nucleoside*
adenine	adenosine
uracil (U) (X = H)	uridine
guanine	guanosine
cytosine	cytidine

Fig. 7.3. Two approaches to dinucleotide synthesis.

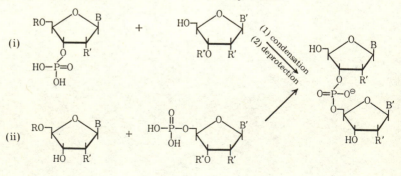

Synthesis of 64 ribotrinucleoside diphosphates by H. G. Khorana (Fig. 7.4) was accomplished by the stepwise addition method, in which 3'-nucleotides were added one by one to 2',3'-protected nucleosides. The fact that these 64 triplets were synthesized by four to five chemists in nearly a year, attests to the excellence of this method. These 64 triplets contributed greatly to the determination of the genetic code.

Repetition of this triplet synthesis leads to large nucleotides by this stepwise addition method. For example, Fig. 7.5 shows the synthesis of a fragment of

Fig. 7.4. Synthesis of the triplet AUG.

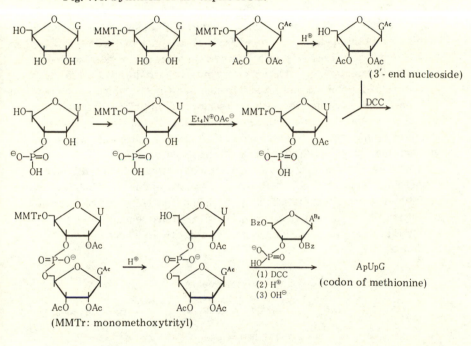

(MMTr: monomethoxytrityl)

Fig. 7.5. Synthesis of tRNA$_f^{Met}$ fragment.

$$HOG^{iBu}(OBz)_2 \xrightarrow[\substack{(1)\ DCC \\ (2)\ H^\oplus}]{MMTrG^{iBu}(OBz)p} HOG^{iBu}(OBz)pG^{iBu}(OBz)_2$$

$$\xrightarrow[\substack{(1)\ DCC \\ (2)\ H^\oplus}]{MMTrC^{Bz}(OBz)p} HOC^{Bz}(OBz)pG^{iBu}(OBz)pG^{iBu}(OBz)_2$$

$$\longrightarrow HOU(OBz)pC^{Bz}(OBz)pG^{iBu}(OBz)pU(OBz)pC^{Bz}(OBz)pG^{iBu}(OBz)pG^{iBu}(OBz)_2$$

$$\xrightarrow[\substack{(1)\ DCC \\ (2)\ H^\oplus \\ (3)\ NH_4/MeOH}]{MMTrG^{iBu}(OBz)p} GpUpCpGpUpCpGpGp \quad \text{(the } 47 \sim 54\text{th fragments of tRNA}_f{}^{Met})$$

E. coli tRNA$_f^{Met}$ 47-54th nucleotides. Use of dicyclohexylcarbodiimide (DCC), which is easy to handle, gives a molecule of the defined sequence. However, if the chain length is elongated, isolation of products from the starting material becomes difficult and incoming mononucleotide units must be used in large excess to maintain good yields at each step. By this method, the largest oligonucleotide obtained was the 12-mer.

Sometimes, instead of an oligonucleotide having no phosphate at either end, a molecule having a phosphate at one end is desired. In such a case, block oligomers can be synthesized for use in the block condensation method. This method compensates for the drawbacks in the above stepwise condensation method. Block condensation is widely used in peptide synthesis and can yield long-chain polypeptides. The systematic studies of Khorana are representative in this area. In this method, two 5'-nucleotides are condensed with one phosphate protected by a β-cyanoethyl group. However, in order to elongate the chain by repeated condensation, a protecting group must be introduced again to the phosphate. This point was improved by using $CCl_3CH_2O^-$, which can be removed with Zn–Cu, an alkylthio group which is removable with alkali after oxidation, and aromatic amidates which can be removed with amyl nitrite at neutrality and room temperature. Using the anilidate for ribopolynucleotide synthesis yielded ribooligonucleotides for block condensation.

If we obtain these medium-size oligomer blocks, we can then synthesize long-chain polynucleotides of defined sequences by joining these blocks. A condensation reaction along these lines has been applied to oligomer blocks containing thymidine, and the first example with a defined sequence was the synthesis of an eicosanucleotide of part of a yeast tRNA^{Ala} gene by Weber & Khorana (Fig. 7.6).

Fig. 7.6. Synthesis of a DNA fragment.

$$MMTrG^{iBu}\text{-OH} \xrightarrow{pA^{Bz}\text{-OAc}} MMTrG^{iBu}pA^{Bz}\text{-OH} \xrightarrow{pA^{Bz}\text{-OAc}} MMTrG^{iBu}pA^{Bz}pA^{Bz}\text{-OH}$$

$$\xrightarrow{pC^{An}pC^{An}\text{-OAc}} MMTrG^{iBu}pA^{Bz}pA^{Bz}pC^{An}pC^{An}\text{-OH}$$

$$\xrightarrow{pG^{iBu}pG^{iBu}pA^{Bz}\text{-OAc}} MMTrG^{iBu}pA^{Bz}pA^{Bz}pC^{An}pC^{An}pG^{iBu}pG^{iBu}pA^{Bz}\text{-OH}$$

$$\xrightarrow{pG^{iBu}pA^{Bz}pC^{An}pT\text{-OAc}} MMTrG^{iBu}pA^{Bz}pA^{Bz}pC^{An}pC^{An}pG^{iBu}pG^{iBu}pA^{Bz}\text{-}$$
$$pG^{iBu}pA^{Bz}pC^{An}pT\text{-OH}$$

$$\xrightarrow{pC^{An}pTpC^{An}\text{-OAc}} MMTrG^{iBu}pA^{Bz}pA^{Bz}pC^{An}pC^{An}pG^{iBu}pG^{iBu}pA^{Bz}\text{-}$$
$$pG^{iBu}pA^{Bz}pC^{An}pTpC^{An}pTpC^{An}\text{-OH}$$

$$\xrightarrow{pC^{An}pA^{Bz}pTpG^{iBu}\text{-OAc}} MMTrG^{iBu}pA^{Bz}pA^{Bz}pC^{An}pC^{An}pG^{iBu}pG^{iBu}pA^{Bz}\text{-}$$
$$pG^{iBu}pA^{Bz}pC^{An}pTpC^{An}pTpC^{An}pC^{An}pA^{Bz}pTpG^{iBu}\text{-OH}$$

$$\xrightarrow{NH_3/MeOH} GpApApCpCpGpGpApGpApCpTpCpTpCpCpApTpG\text{-OH}$$

This is seldom done with the ribo series, but the synthesis of the $3'$-end nonanucleotide of yeast tRNA^{Ala} is one successful example (Fig. 7.7).

7.2.2 *Synthesis by the phosphotriester method*

In the phosphodiester method described above, the oligomers usually contain dissociated phosphates, and for product isolation, hydrophilic solvents and ion-exchange chromatography must be used for the purification. Although in recent years various supports for chromatography have been developed and high pressure liquid chromatography (HPLC) is being utilized, treatment with a large volume of water solution is still a rate-limiting factor.

This problem was fundamentally improved by the triester method, which was first attempted by R. L. Letsinger (Fig. 7.8). The fully protected inter-mediates used in this approach can be isolated by silica gel chromatography and be obtained as powdery solids. These points make this method suitable for large-scale treatment and Itakura used it to synthesize the somatostatin gene (Fig. 7.9).

Fig. 7.7. Block condensation of ribotrinucleotides corresponding to the tRNA^{Ala} $3'$-end.

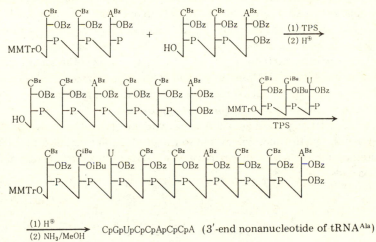

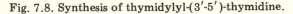

CpGpUpCpCpApCpCpA ($3'$-end nonanucleotide of tRNA^{Ala})

Fig. 7.8. Synthesis of thymidylyl-($3'$-$5'$)-thymidine.

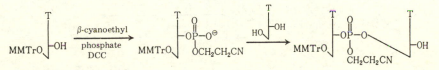

Recently, we used *o*-nitrobenzyl, which is removable by photoirradiation, for
2′-OH protection and *p*-chlorophenylanilidophosphorochloridate for 3′-phos-
phorylation and were able to synthesize decaribonucleotide corresponding to
the 11-20th nucleotides of $tRNA_f^{Met}$ (Fig. 7.10).

The phosphotriester method has merits making it suitable for large-scale
synthesis and should be widely utilized in the future.

7.2.3   *Condensation of deoxyoligonucleotides using enzymes*

As discussed above, chemical synthesis of long-chain polynucleotides
has some limitations, and polynucleotides that are large enough to be useful
for biochemical and genetical research still cannot be obtained. To overcome
this problem, attempts have been made to prepare oligonucleotides of defined
sequences by using the proper enzymes, and some successful results have been
reported.

First was the synthesis of a double-stranded DNA of 77 chain-length corresponding to $tRNA_{Yeast}^{Ala}$ by Khorana *et al.* DNA ligase obtained from T4 phage
infected *E. coli* cells was utilized to join the oligonucleotides. Khorana and
collaborators further succeeded by synthesizing a gene for a suppressor tRNA,
$tRNA_{Su\ III}^{Tyr}$, which had been proved to be biologically active. Fig. 7.11 shows
the oligomers used in this experiment and the order in which they are joined.
The gene obtained was introduced into λ phage and revealed a suppressor
activity.

Using the same method, H. W. Boyer *et al.* synthesized the somatostatin gene,
combined it with the lactose operon and introduced it into plasmids of *E. coli*,
which would then produce somatostatin, a growth hormone inhibitor. Recently,
insulin has been produced by the same approach.

The genetic engineering used in these experiments should contribute to
research in genetics and biology and to the production of medicines and agricultural products, although care must be exercised toward protection against
biohazards.

Fig. 7.9. An example of the triester method.

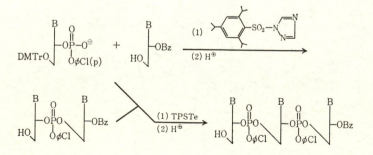

7.3   Applications of synthetic polynucleotides

Combinations of various chemical synthetic methods and enzymic reactions can afford polynucleotides of defined sequences as described above. These polynucleotides can serve genetics and biochemistry in many ways.

7.3.1   *Artificial synthesis of tRNA*

In 1965, Holley *et al.* first determined the whole base sequence of a tRNA and proposed a secondary structure for it – the so-called clover leaf model. This was the first evidence that nucleic acid is a natural product molecule.

tRNA$_f^{Met}$ from *E. coli* not only accepts methionine with the aid of aminoacyl tRNA synthetase, but also interacts with the initiation factor as protein

Fig. 7.10. Triester synthesis of ribodecanucleotide corresponding to tRNA$_f^{Met}$ 11–20 nucleotides.

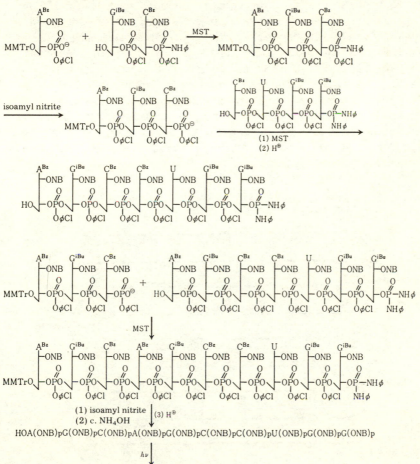

biosynthesis initiation tRNA. Moreover, this tRNA forms a complex with ribosomes at a specific site. This makes it interesting biochemically and suitable for nucleic acid – protein interaction studies. If this initiation reaction can be inhibited, a new compound for inhibiting protein biosynthesis can be formed. Although not directly related to $tRNA_f^{Met}$, a type of tRNA acts as a primer for reverse transcriptase in the synthesis of oncogenic virus DNA. In this case tRNA, also functions as a regulatory compound in DNA synthesis and the mechanism of oncogenicity by these viruses is interesting.

Thus, $tRNA_f^{Met}$ synthesis requires: (i) synthesis of fragments corresponding to each part of the tRNA, and (ii) joining of these fragments by some means. To synthesize the fragments, the di- or triester method described above can be used. Let us consider the synthesis of *E. coli* $tRNA_f^{Met}$, the total structure of which was reported by S. Dube *et al.* (Fig. 7.12). First, we had to decide how to select the tRNA fragments. We chose to synthesize fragments of 5–10 nucleotides, because fragments of this size can be synthesized with some degree of certainty. Using the phosphodiester method, we synthesized fragments 1a (No. 1–4), 1b (No. 5–10), 6 (No. 35–40), and 7 (No. 41–46) as hexamers, and 8 (No. 47–54), 9 (No. 55–57) and 10 (No. 58–60). Next, fragments 11a (No. 61–65) and 11b (No. 66–71) were synthesizes by a combination of the di- and triester methods. Fragments 11c (No. 72–74) and 11d (No. 75–77) were synthesized by adding protected Ap and Cp using partially nitrobenzylated nucleotides. The last fragments 2 (No. 11–20), 3 (No. 21–24), 4 (No. 25–30), and 5 (No. 31–34) were synthesized by the triester approach described above.

Fig. 7.11. Gene for $tRNA_{SuIII}^{Tyr}$ (oligonucleotides between the notches were chemically synthesized).

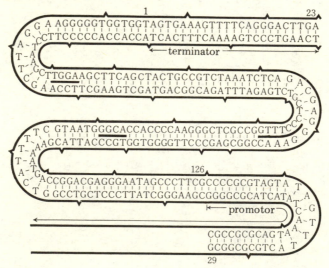

All fragments corresponding to the tRNA base sequences could be synthesized, but joining them was another problem. We considered (i) joining these oligomers with DNA ligase, which has been used in DNA synthesis, and (ii) chemical ligation on the template. Template synthesis has been studied extensively by Orgel *et al.* in relation to the origin of life. The general principle is to bind complementary monomers or oligomers by hydrogen bonding in a double-strand manner and join them using water-soluble carbodiimide. To adopt this method, oligomers corresponding to double strands are required and twice the effort is required. Moreover, stabilization of the complex must be studied further.

RNA ligase was isolated by Hurwitz *et al.* from *E. coli* infected with T4 phage, like DNA ligase, and reported to join poly(A) strand at $3'$-OH and $5'$-end phosphate. Later, it was found that under optimum conditions, this enzyme can join two ribooligonucleotides and is suitable for joining tRNA fragments.

We joined four chemical fragments of tRNA$_f^{Met}$ at the $5'$-end as shown in Fig. 7.13. Joining the first two fragments gave two decamers, which were then joined to obtain an eicosanucleotide CGCGGGGUGGAGCAGCCUGGp (corresponding to 1–20 nucleotides of tRNA$_f^{Met}$). Analogously, the tRNA$_f^{Met}$ $3'$-end heptadecanucleotide UCCGGCCCCCGCAACCA was synthesized. To obtain the total molecule, a $5'$-quarter molecule (No. 1–20) served as an acceptor in the joining reactions with the $3',5'$-bisphosphorylated donor molecule (No. 21–34) to afford the $5'$-half molecule. This was then joined

Fig. 7.12. *E. coli* tRNA$_f^{Met}$ structure. Oligonucleotides between partitioning lines were chemically synthesized. The broken and solid lines indicate the positions at which the RNA ligase joining took place.

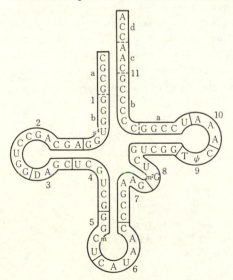

to the 3′- half molecule which was prepared by ligation of the other quarter molecules (No. 35–60, acceptor, No. 61–77, donor) followed by 5′-phosphorylation with polynucleotide kinase. Thus the total molecule of *E. coli* tRNA$_f^{Met}$ was obtained by chemical and enzymic means.

7.3.2   *Factors determining the winding direction of the nucleic acid helix*
Nucleic acids, especially DNA have right-handed helical structures as Watson and Crick postulated. Why is DNA a right-handed helix? As the Watson-Crick model was elucidated from the DNA fibre pattern determined by X-ray crystallography and model building, it does not answer this question. Since nucleic acid is a long molecule formed from nucleosides joined by 3′-5′ phosphodiester linkages, the sugar-phosphate backbone has five freely rotating bonds and the nucleoside structure has two. The angles of these last two are *endo–exo* puckering of the sugar 2′-3′ bond and *syn–anti* rotation of bases around the nucleosidic linkages. Nucleic acid helices are stabilized in the minimum energy position. In addition to this rotational restriction, electrostatic repulsion by phosphate dissociation, hydrophilic interaction of bases (stacking), and – in double helices – complementary hydrogen bondings between bases might be stabilization factors. If the rotation of bases around the nucleosidic bond at a certain angle is stopped and puckering of the sugar is fixed, oligonucleotides and polynucleotides of this type of nucleoside may have interesting structures.

For this purpose, cyclonucleosides with an additional bond between base and sugar and a fixed rotational angle of base (χ) and sugar puckering might be suitable. In oligomers consisting of cyclonucleosides, only the backbone is variable and thus their structure might be studied rather precisely.

Fig. 7.14 gives examples of cyclonucleosides. These nucleosides were obtained as single crystals and the torsion angles (χ) were determined. We synthesized A^{s}pA^s in which two 8,2′-S-cycloadenosines were joined by a phosphodiester linkage between 3′ and 5′ hydroxyls. The circular dichroic (CD) spectrum of A^{s}pA^s, shown in Fig. 7.15(*a*), has a CD curve with (−)-(+) Cotton bands from the long wavelength side, an inverted form of the ApA spectrum. Tinoco's

Fig. 7.13. RNA ligase joining of chemical fragments.

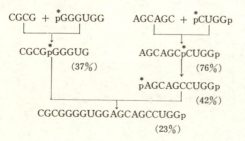

study showed that these bands are caused by splitting of two transition moments of two overlapped bases, and if the bases overlap in right-handed fashion, it is $(+)-(-)$ and if in left-handed, it is $(-)-(+)$. According to this theory, ApA has right-handed stacking like in natural RNA and $A^s pA^s$ or $A^o pA^o$ has reversed left-handed stacking. Oligonucleotides produced by polymerization of pA^s also had the same splitting band of the $(-)-(+)$ type and this property was maintained in the longer chain.

In order to confirm this point, ^1H NMR of $A^s pA^s$ was examined. Signals corresponding to H-1$'$ (sugar) and H-2 (base) showed shifts as expected in left-handed stacking. Each signal was confirmed by a broadening experiment with

Fig. 7.14. Structures of 8-cyclonucleosides.

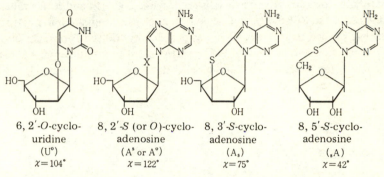

6, 2$'$-O-cyclo-
uridine
(U^o)
$\chi = 104°$

8, 2$'$-S (or O)-cyclo-
adenosine
(A^s or A^o)
$\chi = 122°$

8, 3$'$-S-cyclo-
adenosine
(A_s)
$\chi = 75°$

8, 5$'$-S-cyclo-
adenosine
($_sA$)
$\chi = 42°$

Fig. 7.15. (a) CD spectra of $A^s(pA^s)_n$, (b) UV and CD spectra of $A^o pA^o$ (——) and ApA ($\cdots$).

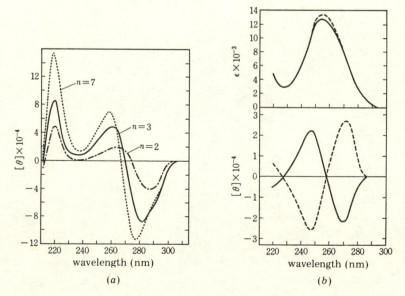

$M_n^{2\oplus}$. These results can be explained well if left-handed stacking is assumed for these oligomers.

An energy calculation of A^spA^s in which the angles around each axis of rotation were minimized also showed this left-handed structure. Fig. 7.16 presents this optimum structure obtained by computer calculation of A^spA^s in contrast to the natural RNA right-handed helical structure.

Double-stranded complexes from these cyclonucleotide oligomers with poly(U) of a right-handed nature were not formed. An oligomer poly(U°), which has uracil bases complementary to adenine and a χ value almost the same as that of poly(A°), was synthesized. This oligomer hybridized with oligo(A^s) to form a double helix.

These studies suggest that the most important factor defining the direction of rotation of nucleic acid helices is the torsion angle (χ) and if this value is $100 \sim 200^\circ$ (so-called *high–anti* region) in nucleoside units in polynucleotides, the left-handed structure is taken in single and hybridized forms. On the other hand, when in natural nucleic acids χ of the constituent nucleosides is between 0 and 30° (*anti* region), the right-handed structure is easily assumed.

Fig. 7.16. Left-handed helical structure of an A^s oligomer.

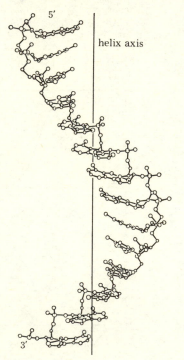

7.3.3 Biological activity of polynucleotides

Nucleoside 5'-diphosphates are easily polymerized by polynucleotide phosphorylase. From natural nucleoside diphosphates, ADP, UDP, CDP, GDP and IDP, poly(A), poly(U), poly(C), poly(G) and poly(I) were synthesized respectively, and their physical properties and biological activities were studied. As the double-stranded complex, poly(I) · poly(C), has interferon-inducing activity, nucleoside analogue diphosphates have been synthesized and polymerized to study this activity. Recently, these polynucleotides have been discussed as inhibitors against (reverse) transcriptase. The enzyme is specifically important in the multiplication of mammalian oncogenic viruses, and therefore these polynucleotides may be interesting to consider as anti-cancer agents.

(i) Interferon-inducing activity. Several polynucleotides studied as potential inducers are listed in Table 7.1. Clearly, only double-stranded polynucleotides showed the activity, with single- and triple-stranded ones having no activity. For double-strandedness, both chains need complementary bases. A–U or I–C analogues were well investigated. At present, poly(I) · poly(C) is the strongest inducer, and studies related to this compound are accumulating. Thus, the interferon-inducing property seems to require the chain length to be long enough ($\sim$100 bases) at least in one strand, T_m to be higher than the usual temperature, and both strands to have 2'-OH. The only one equivalent to poly(I) · poly(C) was a thiophosphate polymer.

Various groups can now be introduced to the 2'-position of purine nucleosides by breaking the cyclonucleoside. Purine nucleoside diphosphates having N_3, NH_2 and halogens at the 2'- position have been synthesized, and various poly(dA) and poly(dI) analogues have been obtained. As shown in Table 7.1, poly(dIz) · poly(C) and poly(dIf) · poly(C) showed activity as high as that of poly(I) · poly(C), and poly(dIcl) · poly(C) also had high activity. However, to our surprise, poly(dIz) · poly(br^{5}C) had no activity in spite of its high T_m. These results show that 2'-OH does not have to be present in both strands. This indicates that not merely the 2'-OH but the overall tertiary conformation is recognized by the interferon-producing system.

(ii) Inhibition of reverse transcriptase. Reverse transcriptase follows the basic concept of biology of DNA $\rightarrow$ RNA $\rightarrow$ protein and catalyzes RNA $\rightarrow$ DNA synthesis. Moreover, this enzyme participates in the multiplication of oncogenic viruses.

This enzyme acts, as illustrated in Fig. 7.17, as a catalyst of the synthesis of complementary DNA on RNA using certain tRNAs as primers. Synthetic oligonucleotides inhibit this DNA synthesis in some steps. Table 7.1 (right column) shows the inhibitory activities of polynucleotides. Poly(ms^{2}A) and poly(ms^{2}I) with the thioalkyl group at the purine position-2 are active inhibitors, suggesting that these polymers interact with enzyme active sites.

7.4 Completely artificial information-transmitting macromolecules

As discussed in the beginning of this chapter, some information-transmitting macromolecules (if the information is very simple) seem to exist, and for practical purposes synthesis of completely artificial molecules may be important. The reasoning behind this may be likened to the replacement of silk or linen by synthetic fibres, where the basic properties are retained but the strength is increased.

Table 7.1. *Interferon-inducing activity of various polynucleotides*

	Interferon-inducing activity	Reverse transcriptase-inhibiting activity
Single strand		
poly(I)	−	+
poly(C)	−	+
poly(ms^{2}A)		++
poly(es^{2}A)		++
poly(ms^{2}I)	−	+
poly(C^{7}A)	−	+
poly(C^{7}I)	−	+
poly(C^{3}A)	−	−
poly(C^{3}I)	−	−
poly(n^{2}A)	−	+
poly(n^{2}I)	−	+
poly(dAz)	−	++
poly(dAf)		(Enhanced activity)
Double strand		
poly(A) · poly(U)	+	−
poly(I) · poly(C)	++	−
poly(I) · poly(dCz)	−	−
poly(I) · poly(br^{5}C)	++	−
poly(C^{7}A) · poly(U)	−	−
poly(C^{7}I) · poly(C)	+	−
poly(C^{3}A) · poly(U)	−	−
poly(C^{3}I) · poly(C)	−	−
poly(n^{2}A) · poly(U)	−	−
poly(n^{2}I) · poly(C)	−	−
poly(dAz) · poly(U)	−	−
poly(dAz) · poly(rT)	−	−
poly(dAz) · poly(dUz)	−	−
poly(dIz) · poly(C)	++	−
poly(dIz) · poly(br^{5}C)	−	−
poly(dIcl) · poly(C)	+	−
poly(dIf) · poly(C)	++	−

Biopolymers are not designed for only a few limited functions but for many purposes. They are synthesized from raw materials readily available in biological systems; they are recognized by organs or tissues; they can be transferred from the place of synthesis to the place of functioning, and are degraded when they become useless. Thus they contain not only the required structure units, but also those useless to certain functions. The aim in creating artificial macromolecules is to have them transmit only the information useful or required for a certain purpose.

Very simple, information-transmitting macromolecules can be expressed as in Fig. 7.18. Since information is transmitted from molecule to molecule, the single interactive unit (or information unit) may possess minimum elementary information. If this is given in sequence, information of higher grade will be given according to the length (to a receiver having recognition ability). When one information unit is 'planted' on macromolecules, one functional macromolecule transmits only a single piece of information. This is the most easily applicable concept and is illustrated by the following example.

Takemoto *et al.* succeeded in separating thymine and adenine from a mixture by absorption column chromatography on poly(1-vinyluracil) resin [1]. In this case, the polymer has only one piece of information: to recognize adenine using uracil. This type of resin can be used for base separations in general. For a simple

Fig. 7.17. Schematic representation of the action of reverse transcriptase.

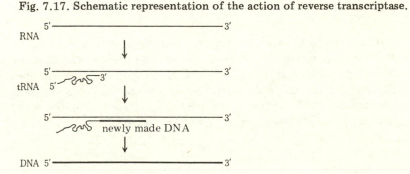

Fig. 7.18. Model of an information-transmitting macromolecule.
Information unit
Sequence of information, total number 2
Sequence of information, total number n.

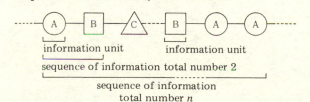

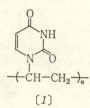

[1]

function like this, a sequence of information is not required, but if we utilize even a homopolymer of certain nucleic acid bases for the separation, the selection becomes the 'one out of four' type and not the 'one from two'.

Next, let us consider an example for simple sequential information (Fig. 7.19). Preparation of meaningful and useful sequential information is difficult, mostly due to the lack of efficient synthetic technique. At present, only a repeated number of n can be transmitted. Naylor reported that coupling of hexathymidate on poly(A) is accelerated by about tenfold to give T_{12} and if adenylate is present in this system, A_{12} is also formed by a 'template effect', i.e. the product is directed by the template. This fact may be interpreted as:

$$(pT)_5 pT\text{—OH} + p\text{—T}(pT)_5 \xrightarrow[\substack{\text{condensing} \\ \text{agent}}]{A_n} T_{12} \overset{\substack{12 \times pA \\ \text{condensing reagent}}}{\underset{\substack{\text{condensing} \\ \text{reagent}}}{\diagdown}} A_{12} \text{—} 12 \times pT$$

$$(\equiv T_6) \qquad (\equiv T_6)$$

In this reaction, recognition between molecules is not satisfactory (in solution) and accurate information transfer should not be expected. Therefore, the n value of the product is distributed centering at 12. In this case, however, information-transmitting macromolecules are repeatedly used as catalysts and replicate the information. This is very important and the polymerization degree n and the base (one out of four) information are effectively transferred.

Fig. 7.19. Completely artificial information-transmitting polymer.

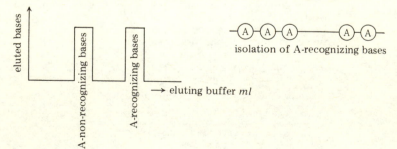

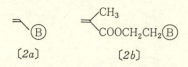

[2a]   [2b]

Many similar experiments followed these famous experiments of Naylor. Takemoto *et al.*, studied mutual interaction in solution and used the *N*-vinyl derivative of nucleobases [2a] or *N*-β-metacrylate oxyethylester [2b] in vinyl-polymerization. When A and U were used as ⑧, the rate of polymerization was maximum in a 1:1 ratio, and in an optimum solvent system a certain amount of altering polymerization took place. If we develop this system, we may be able to synthesize a macromolecule like $+$Ⓐ$-$Ⓑ$+_n$ with quite high information transfer. Also, if we utilize [3] as an information-transmitting macromolecule, polymerization of a nucleobase monomer, which makes pairing with base ⑧, would be accelerated especially for the isotactic isomer.

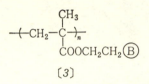

[3]

Ise *et al.* investigated this problem in more detail by synthesizing a macromolecule having nucleobases and cations such as [4]. The interactions of nucleobase polymers themselves or with DNA were investigated by the absorbance change in the electronic spectrum. Together with hydrogen bonding, hydrophobic and electrostatic interactions of bases are also important. A graph with the molar percentage of the nucleobases on the ordinate and the absorbance on the abscissa shows the mole percentage of specific interaction as a sharply bent curve. Ise *et al.* also synthesized a resin-type information-transmitting polymer from isoamylpyridine and styrene, where base stacking was mainly observed as reported by Shimidzu *et al.* who used a synthetic polymer [5] and found that the interaction decreased in the order of purine–purine > purine–pyrimidine > pyrimidine–pyrimidine. Shimidzu studied the 'template' effect for [6] and [7]. If Ⓟ-TCTC (Ⓟ is polymer) is used as the template, two trinucleotides pdApdGpdA and pdGpdApdG are obtained together with various dinucleotides. Therefore hydrogen bonding functions effectively as an information-transmitting interaction in three units of nucleotides, and a shorter information source is not effectively transmitted.

These studies on information-transmitting polymers are still at the initial stages and only low-grade information can be transmitted, mainly because of the lack of well-organized environments around nucleobases. In solution there is a strict limitation to the transfer of information simply via hydrogen bonding since the recognition interaction of simple hydrogen bonding in solution is weak. Fortunately, the study of nucleic acids makes it evident that a specific microenvironment of nucleobases exists, based on evidence from a sharp

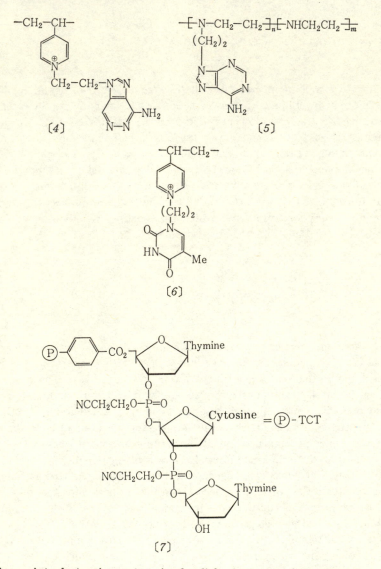

[4]

[5]

[6]

[7]

transition point, electronic spectra, circular dichroic spectra, fluorescence spectra, and NMR. By exploiting this information and applying new synthetic techniques, we should be able to obtain very useful, artificial macromolecules that transmit high-grade information.

FURTHER READING

CHAPTER 2

A few reference books were published as a general guide to the reactions of polymers, while the individual items have been reviewed; for instance, in *Macromolecular Reviews* (Interscience USA), *J. Macromol. Sci.*, *Rev. Macromol. Chem.* (Marcel Dekker, USA).

F. M. Fettes, ed., *Chemical Reaction of Polymers*, Interscience USA, 1964. The first book to summarize polymer reactions. The sixteen sections, each written by a specialist, include the reactions of individual polymers, such as unsaturated polymer, cellulose, etc., in the first half of the book and decomposition, coupling, oxidation, etc., in the second half of the book.

M. Okawara, *Chemical Reactions of Polymers*, vol. 1, *Polymer Reaction*, Kagaku Dōjin, Japan, 1972 (in Japanese). This volume includes the reactions of individual polymers, intramolecular reactions, graft and block polymerization, intermolecular reactions of polymers, decomposition, and polymer effects. Also available from the same publisher is vol. 2 of the series: *Functional Polymers*.

J. A. Moore, ed., *Reactions on Polymers*, D. Reidel, Holland, 1973. This is a collection of papers presented at the symposium held by NATO Scientific Division held at Troy, USA in 1973, which includes specific aspects of polymer reaction, reactions in gel, solid and crystalline states, photoreactions, bio-related reactions, catalytic reactions and so on.

C. G. Overberger & B. Sedláček, eds., *Transformations of Functional Groups on Polymers*, Interscience, USA, 1974. Collective papers presented to the 13th Microsymposium (IUPAC) held at Prague, Czechoslovakia.

Polymer Reaction, ed. the Society of Polymer Science, Japan, Kyōritsu Shuppan, Japan, 1978 (in Japanese). *Experimental Polymer Science*, vol. 6. Published in thirty years commemoration of the Society of Polymer Science, Japan.

CHAPTER 3A

CATALYSIS BY IONIC AND NON-IONIC MACROMOLECULES

I. Sakurada, *Pure Appl. Chem.*, 16, 263 (1968).

H. Morawetz, *Acc. Chem. Res.*, 3, 354 (1970).

N. Ise, *Adv. Polymer Sci.*, 7, 536 (1970/71).

N. Ise, *Journal of Polymer Science*, Polymer Symposia, 62, 205 (1978).

N. Ise, *Makromol. Chem.*, Suppl. 5, 102 (1981).

N. Ise, T. Okubo & S. Kunugi, *Acc. Chem. Res.*, 15, 171 (1982).

C. G. Overberger & J. C. Salamone, *Acc. Chem. Res.*, 2, 217 (1969).

J. Fendler & E. Fendler, *Catalysis in Micellar and Macromolecular Systems*, Academic Press, New York (1969).

H. Morawetz, *Macromolecules in Solution*, 2nd edn chapter 9. Interscience, New York (1975).

K. Takemoto, T. Kunitake, Y. Imanishi & T. Shimizu, (1976). *Kobunshi-shokubai* (Macromolecular Catalysis). Kodansha, Tokyo.

CATALYSIS BY POLYPEPTIDES

M. Hatano & T. Nozawa, *Prog. Polymer Sci. Japan*, 4, 223.

NUCLEIC ACID MODELS

K. Takemoto, *J. Macromol. Sci.*, Part C 5, 29 (1970).

K. Takemoto, *J. Polymer Sci.*, Polymer Symposia, 55, 105 (1976).

T. Okubo & N. Ise, *Adv. Polymer Sci.*, 25, 136 (1977).

POLYMER-METAL COMPLEXES

H. Tanaka, S. Nakahara & S. Fukui, (1976). *Bioinorganic Chemistry*, chapter 2, Kagaku Dojin, Kyoto.

E. Tsuchida & H. Nishide, *Adv. Polymer Sci.*, 24, 2 (1977).

T. Saegusa, E. Tsuchida & Hirai, H. *Kobunchi Kinzokusakutai-Gosei, Kino, Ōyo* (Polymer-Metal Complex – Synthesis, Functions and Applications), Kagaku Dojin, Kyoto.

HETEROGENEOUS CATALYSIS

Y. Chauvin, D. Commereuc & F. Dawans, *Prog. Polymer Sci.*, 5, 96 (1977).

CHAPTER 3B

REVIEW AND BOOKS

S. Kukui, I. Tabushi & T. Kunitake, *Bioorganic Chemistry*, Kodansha Scientific, Tokyo (1976).

P. D. Boyer, ed., *The Enzymes*, 3rd edn, Academic Press, New York.

E. Zeffren & P. L. Hall, *The Study of Enzyme Mechanism*, John Wiley & Sons, New York (1973).

D. Dolphin, C. McKenna, Y. Murakami & I. Tabushi, *Biomimetic Chemistry*, Advances in Chemistry Series No. 191, American Chemical Society, Washington, DC, (1980).

K. Takemoto & I. Tabushi, *Medical Polymers*, Kodansha Scientific, Tokyo (1978).

W. P. Weber & G. W. Gokel, *Phase Transfer Catalysis in Organic Synthesis*, Springer-Verlag, Berlin (1977).

ORIGINAL PAPERS

I. Tabushi & M. Funakura, *J. Amer. Chem. Soc.*, 98, 4684 (1976). On artificial electron transport membranes.

G. Wulff, W. Vesper, R. Grobe-Einsler & A. Sarhan, *Makromol. Chem.*, 178, 2799 (1977). On specific binding polymers.

CHAPTER 4

ENERGY CONVERSION IN GENERAL

T. Ohta, *Basis of Energy Conversion*, Maki Shoten, Tokyo (1965). This book treats various physical methods of energy conversion. The excellent introductory chapter provides a concise view of energy conversion in general.

Solar Energy. A monthly journal from Pergamon Press, Oxford. Original articles as well as review articles presented at international conferences relevant to solar energy.

Energy and Resources. A bimonthly journal of the Japan Society of Energy and Resources. This journal covers science, technology, economy, and politics concerning to energy and resources.

PHOTOCHEMISTRY

N. J. Turro, *Modern Molecular Photochemistry*, Benjamin, New York (1978). A well-written book of organic photochemistry with a qualitative but profound description of the physical background.

The Chemical Society, *Photochemistry*, Special Periodical Report. An annual report on photochemistry and a valuable literature source in all aspects of photochemistry.

ENERGY CONVERSION IN BIOLOGICAL SYSTEMS

J. Barber, ed., *Photosynthesis in Relation to Model Systems*, Elsevier, Amsterdam (1979).

CHAPTER 5

MEMBRANES, GENERAL

N. Lakshminarayanaiah, *Transport Phenomena in Membranes*, New York, Academic Press (1969).

P. Meares, *Membrane Separation Processes*, Elsevier Scientific, Amsterdam (1976).

M. Senō, *Functions of Membranes (Maku no Kinoo)*, Kyoritsu Shuppan, Tokyo (1977).

M. Nakagaki, *Introduction to Membrane Science (Makugaku Nyumon)*, Kitami Shobo, Tokyo (1978).

S. G. Shultz, *Basic Principles of Membrane Transport*, Cambridge University Press, London (1980).

MEMBRANES, BIOLOGICAL

J. F. Danielli & F. Davson, A contribution to the theory of permeability of thin films, *J. Cell. Comp. Physiol.*, 5, 495 (1935).

S. J. Singer & G. L. Nicolson, The fluid mosaic model of the structure of cell membranes, *Science*, 175, 720 (1972).

R. Sato, Biological membranes as molecular assemblies. In *Physical and Chemical Basis of Life (Iwanami Koza, Gendai Seibutzu Kagaku, Vol. 1)*, ed. T. Ooi & R. Sato, pp. 295. Iwanami Shoten, Tokyo (1975).

Y. Kagawa, *Biological Membranes (Seitai Maku)*, Iwanami Shoten, Tokyo (1978).

TRANSPORT MECHANISM

A. Kotyk & K. Janacek, *Cell Membrane Transport–Principle and Technique*, Plenum, New York & London (1970).

M. Nakao, *Active Transport* (Selected Papers in Biology, Vol. 9). Tokyo University Press, Tokyo (1972).

A. L. Lehninger, *Biochemistry – The Molecular Basis of Cell Structure and Function*, Worth Publishers, New York.

Yu. A. Ovchinikov, V. T. Ivanov & A. M. Shkvob, *Membrane-active Complexones*, BBA Library Vol. 12. Elsevier Scientific, Amsterdam (1974).

Y. Anraku, Membrane transport and metabolic regulation. In *Metabolism and Its Regulation (Iwanami Koza, Gendai Seibutzu Kagaku Vol. 5)*, ed. R. Sato & Y. Nishizuka, p. 17. Iwanami Shoten, Tokyo (1975).

CHAPTER 6

T. Kunitake & Y. Okahata, A totally synthetic bilayer membrane, *J. Amer. Chem. Soc.*, 99, 3860 (1977). The first example of the totally synthetic bilayer membrane is reported.

J. H. Fendler, Surfactant vesicles as membrane mimetic agents: characterization and utilization, *Acc. Chem. Res.*, 13, 7 (1980). Physicochemical characterization of some synthetic bilayer membranes and their application to photochemical energy conversion are discussed.

T. Kunitake & S. Shinkai, Catalysis by micelles, membranes and other aqueous aggregates as models of enzyme action, *Adv. Phys. Org. Chem.*, 17, 435 (1980). The catalytic actions of the ammonium bilayer membrane are summarized.

L. Gros, H. Ringsdorf & H. Schupp, Polymeric antitumor agents on a molecular and on a cellular level? *Angew. Chem. Int. Ed. Engl.*, 4, 305 (1981). Formation and some properties of polymerized bilayer vesicles are discussed.

J. H. Fendler, Membrane mimetic chemistry, Wiley-Interscience, New York, 1982. Characterization and applications of micelles, microemulsions, monolayers, bilayers, vesicles, host-guest systems, and polyions.

CHAPTER 7

M. Calvin, *Chemical Evolution*, Oxford University Press (1969).

W. Guschlbauer, *Nucleic Acid Structure*, Springer, Berlin (1976).

M. Ikehara, E. Ohtsuka & A. F. Markham, The synthesis of polynucleotides, *Adv. Carbohyd. Chem. Biochem.*, **36**, 135 (1979).

E. Ohtsuka, M. Ikehara & D. Söll, Recent developments in the chemical synthesis of polynucleotides, *Nucleic Acids Res.*, **10**, 6553 (1982).

INDEX

A23187 165
absorption column chromatography 213
actin (muscle) 114
actins (macrotetrolides) 165
activated acyl structure 13
activation
    of hydroxyl group 7
    of polymer 6
    of substrates 97
activation parameters 56, 57, 58, 59, 69, 70
active transport 162, 167, 169
ADP → ATP conversion 118
aggregate morphology 179
allosteric enzymes 101
aminoacyl tRNA synthetase 205
aminomethylated polystyrene 23
antamanide 164
antenna pigments 117
apohost 94
artificial
    active site 100
    information-transmitting
        macromolecules 212
    macromolecules 198
    plant 106
ATPase 169, 171
azobenzene 188
azobenzene cluster 193

bilayer 175
block condensation method 202
Brønstead-Bjerrum equation 59
building-up principle 94

channel 165, 166
carboxypeptidase 98
carrier 158. 167, 168
carrier-mediated transport 167
cascade control mechanism 103
catalytic hydrolysis 189, 195
cavities 198
cavity in the polymer 100

CD curve 208
charge-relay system 87
charged membrane 150
chemi-osmotic coupling 162, 170
chemi-osmotic hypothesis 170
chemical energy → mechanical energy
        conversion 119
chiral amphiphile 185
chloromethyl styrene 20
chloromethylated polystyrene 20
chloroplast 115
cholic acid ester 189
cholesterol 189
chromophore stacking 188
chymotrypsin, structure and catalysis 87
circular dichroism 186
colloid osmotic pressure 155
combining protein 169
composite membrane 156
condensation reagents 200
conformation change 87
Coulombic interaction 127
crown ether 160
cupric ion binding 89
cyclam 90
cyclic tetramine 90
cyclodextrin 93
cyclonucleoside 208
cytochrome P-450 99

decarboxylation 191
diacetylene 184
dialkyl amphiphile 176
diallyldiakylammonium chloride,
        copolymers with SO_2 52, 54, 58, 67
diazobenzene 126
DNA ligase 204
double chain ammonium salts 176

$E.\ coli$ tRNA$_f^{Met}$ 202
efficient turnover 104
electro-osmotic pressure 155